Industrial Control Systems

Scrivener Publishing
100 Cummings Center, Suite 541J
Beverly, MA 01915-6106

Publishers at Scrivener
Martin Scrivener (martin@scrivenerpublishing.com)
Phillip Carmical (pcarmical@scrivenerpublishing.com)

Industrial Control Systems

Edited by
Vipin Chandra Pal
Suman Lata Tripathi
and
Souvik Ganguli

Scrivener Publishing

WILEY

This edition first published 2024 by John Wiley & Sons, Inc., 111 River Street, Hoboken, NJ 07030, USA and Scrivener Publishing LLC, 100 Cummings Center, Suite 541J, Beverly, MA 01915, USA

For more information about Scrivener publications please visit www.scrivenerpublishing.com.

Wiley Global Headquarters
111 River Street, Hoboken, NJ 07030, USA

For details of our global editorial offices, customer services, and more information about Wiley products visit us at www.wiley.com.

Library of Congress Cataloging-in-Publication Data

ISBN 9781119829256

Front cover images supplied by Pixabay.com
Cover design by Russell Richardson

Set in size of 11pt and Minion Pro by Manila Typesetting Company, Makati, Philippines

Printed in the USA

10 9 8 7 6 5 4 3 2 1

Contents

Part 2 Control Strategies for Practical Systems 181

List of Contributors

Nitendra Tiwari
Department of Electronics & Instrumentation Engineering,
National Institute of Technology Silchar,
Assam, India

Subhajit Bhattacharya
Department of Electronics & Instrumentation Engineering,
National Institute of Technology Silchar,
Assam, India

Vipin Chandra Pal
Department of Electronics & Instrumentation Engineering,
National Institute of Technology Silchar,
Assam, India

Sudipta Chakraborty
Department of Electronics & Instrumentation Engineering,
National Institute of Technology Silchar,
Assam, India

Sheetla Prasad
Department of Electrical, Electronics & Communication Engineering,
Galgotia University, Greater Noida - Uttar Pradesh, India

G. Boopathi Raja
Department of ECE,
Velalar College of Engineering and Technology, Erode, Tamil Nadu, India

Naiwrita Dey
Department of ECE,
RCC Institute of Information Technology, Kolkata, West Bengal Kolkata, India

Ujjwal Mondal
Department of Applied Physics,
Calcutta University, Kolkata, India

Anindita Sengupta
Department of EE,
IIEST, Shibpur, India
M. Kalpana
Department of ECE,
Kalasalingam Academy of Research and Education, Krishnankovil, Virudhunagar (Dt), Tamil Nadu, India
Krishnapriya J.
Department of ECE,
Kalasalingam Academy of Research and Education, Krishnankovil, Virudhunagar (Dt), Tamil Nadu, India
Sathya Pradeep C.
Department of ECE,
Kalasalingam Academy of Research and Education, Krishnankovil, Virudhunagar (Dt), Tamil Nadu, India
Gayatri P.
Department of ECE,
Kalasalingam Academy of Research and Education, Krishnankovil, Virudhunagar (Dt), Tamil Nadu, India
Santanu Mallick
Department of Applied Electronics & Instrumentation Engineering,
Bankura Unnayani Institute of Engineering, Bankura, India
Ujjwal Mondal
Instrumentation Engineering, Department of Applied Physics,
University of Calcutta, Kolkata, India
R. Ramya
SRM Institute of Science and Technology, Kattankulathur, Chennai, India
M.V. Suganyadevi
Saranathan College of Engineering, Trichy, India
S. Usha
SRM Institute of Science and Technology, Kattankulathur, Chennai, India
Souvik Ganguli
Department of Electrical and Instrumentation Engineering,
Thapar Institute of Engineering and Technology, Patiala, Punjab, India
Yashinidhi Srivastva
Department of Electrical and Instrumentation Engineering,
Thapar Institute of Engineering and Technology, Patiala, Punjab, India
Gagandeep Kaur
Department of Electrical and Instrumentation Engineering,
Thapar Institute of Engineering and Technology, Patiala, Punjab, India

Prasanta Sarkar
Electrical Engineering Department,
National Institute of Technical Teachers' Training & Research, Kolkata, West Bengal, India
Roop Pahuja
Dept. of Instrumentation & Control Engineering,
Dr. B R Ambedkar National Institute of Technology, Jalandhar (Punjab) India
Ranveer Singh
Dept. of Instrumentation & Control Engineering,
Dr. B R Ambedkar National Institute of Technology, Jalandhar (Punjab) India
D. Stalin David
Department of CSE,
IFET College of Engineering, Villupuram, Tamil Nadu, India
Sayan Das
Department of AEIE,
RCC Institute of Information Technology, Kolkata, West Bengal, India
Naiwrita Dey
Department of ECE,
RCC Institute of Information Technology, Kolkata, West Bengal, India
J. Kundu
Department of Electrical Engineering, UCET, Bikaner, Rajasthan, India
A. Choumal
Department of Electrical Engineering, UCET, Bikaner, Rajasthan, India
Sandhya Kumari
Department of Electrical Engineering, Indian Institute of Technology (ISM), Dhanbad, Jharkhand, India
Gauri Shankar
Department of Electrical Engineering, Indian Institute of Technology (ISM), Dhanbad, Jharkhand, India
S. Sriram
Dept. of Computer Engineering,
Sri Ramakrishna Polytechnic. College, Coimbatore, Tamil Nadu, India

Preface

Human civilization started by controlling natural parameters, like fire for making food, water for irrigation to produce grains, creating housing and other products out of wood, etc. Products have turned into a daily need for the people. Controlling input parameters leads to the advancement of manufacturing and other industries. Therefore, advancement of control theory is always a primary focus of research in the development of industry. In view of modern industries, the journey of industrial control systems may be viewed from the control of steam engine to spacecraft, airplane missile control systems to networked control systems, and cybersecurity controls in the new emerging field of robotics. More specifically, in terms of controllers, the journey of the industrial revolution started from the design of P-I-D controllers, fuzzy controllers, neuro-fuzzy controllers, back stepping controllers, sliding mode controllers, and event-triggered controls for networked control systems. Recently, control theory has spread its golden wings in different fields of engineering by use of splendid tools of the control system. In this era, the boom of Internet of Things is exploding at a maximum pace. Different biomedical applications also come under this umbrella and provide the easiest way for continuous monitoring. Some of the prominent research areas in green energy and sustainable development in which control is playing a vital role is as a load frequency controller to control solar thermo plants, event-driven building energy management systems, speed-sensorless voltage and frequency control in autonomous DFIG-based wind energy, hazardous energy control programs, and many more. This book aims to cover all of these subjects and more, in our push to further the field of industrial control systems. We hope this is useful for engineers, scientists, industry professionals, students, and faculty at all levels.

Organization of Book

This book is composed of two parts which contain 13 chapters. Part 1 begins with an introduction to different control techniques which are used in the industrial revolution for making modernization in industry. Chapter 1 describes the history of industry and also classifies the different categories of industries. The primary concerns of working in industry are safety and precautions. The safety monitoring and control of an industrial boiler is explained in Chapter 2 for preventing major accidents in the plant. Chapter 3 reviews the existing literature in controller design for robust control of rotary machineries involved in many industrial applications which are often subjected to position dependent disturbances. A modified robust approach of finite dimensional repetitive controller design and its synthesis is proposed here. After safety, security is a major concern of many precious machinery and other equipment which are used in real time application in Chapter 4. Chapter 5 proposes about advanced adaptive control applied to nonlinear plants. In Chapter 6, the discussion of optimization algorithm PSO and SVM is included and it is analyzed that they are responsible for identifying the appropriate controls for various operating conditions. Two popular algorithms, viz. the firefly technique abbreviated as FA and the pattern search (PS) technique, are combined to develop a new topology termed the hybrid firefly algorithm, designated in Chapter 7 as the FAPS method. Motivated to re-design some of the essential subsystems of a DC micro-grid for efficient power management in a more informative and user-friendly manner, Chapter 8 represents a virtual instrumentation metering infrastructure and a modern ICS (Industrial Control System) for the same.

Part 2 starts with the real time implementation of different control strategies to practical systems.

Chapter 9 proposes an Adaptive Fuzzy Optimal Power Controller (AFOPC) for speed regulation of Brushless Direct Current Motors (BLDCM) and by using this technique, it is feasible to achieve the lowest possible losses while maintaining the highest possible motor speed. New industrial development is going into new dimensions by including robots for different manufacturing industries where humans are not able to work. In Chapter 10, a Fuzzy logic based partially-optimized FOPID controller is developed mainly focusing on controlling robot manipulators. Chapter 11 represents a brief study, analysis, validation, and control of an electromagnetic attraction type levitation set-up where a flat plate has been successfully levitated at a desired operating point. The final product of any industry is used by consumers worldwide and petrol/diesel based transport is a common facility for receiving items by customers. To protect the environment from pollution, the automobile industry is shifting towards the electric vehicle. In Chapter 12, two-area MSIPS with nonlinearities is attempted to examine the effectiveness of a proposed SSA tuned CTI-TID controller on electric vehicles. The modern era is the transfer of information through a single click via different networks. Therefore, security and control are very important and from here the new emerging field of Cyber Security Control Systems is gaining prime importance. In Chapter 13, cyber security control systems for operational technology have been discussed.

Part 1

ADVANCED CONTROL TECHNIQUES

1

Introduction: Industrial Control System

Nitendra Tiwari[1*], Subhajit Bhattacharya[1], Vipin Chandra Pal[1], Sudipta Chakraborty[1] and Sheetla Prasad[2]

[1]Department of Electronics and Instrumentation Engineering, National Institute of Technology Silchar, Assam, India
[2]Department of Electrical Engineering, Galgotias University, Noida, Uttar Pradesh Punjab, India

Abstract
In this chapter, the meaning of industry and the types have been discussed. The history of industry and their way of working are described to focus on the beginning of industry. Finally, the future of industries has been discussed with consideration to the new recent technology in different fields like communication, networking, etc.

Keywords: Industry, revolution of industry, classification of industries

An industry is a collection of firms that have common basic business operations. In today's economy, there are hundreds of industry categories. Typically, industrial groupings are grouped together into larger groups called sectors.

Individual firms are classified into industries based on their principal sources of revenue. Despite the fact that a vehicle manufacturer's finance sector may account for 10% of total sales, most categorization methodologies identify the firm as part of the automobile industry.

**Corresponding author*: nitintiwari22@gmail.com

Vipin Chandra Pal, Suman Lata Tripathi, and Souvik Ganguli (eds.) Industrial Control Systems, (3–14) © 2024 Scrivener Publishing LLC

1.1 Types of Industry

On the basis of raw materials and their processing, industries can be divided as follows.

1.1.1 The Primary Sector

Forestry, agriculture, fishing, quarrying, mining, and mineral exploitation are all part of a country's economy. It can be divided into two categories:

- **Genetic Industry** - This category comprises any raw material manufacturing that can benefit from human involvement in the development process.
- **Extractive Industries** - This category involves the development of finite raw materials that cannot be replenished by cultivation.

The primary industry continues to dominate the economies of underdeveloped and growing countries, but as the secondary and tertiary industries increase, its share of total economic output decreases.

1.1.2 The Secondary Sector

The manufacturing industry (secondary industry) processes the raw materials given by primary industries and converts them into consumer goods. Products that have been included into product lines by specialised secondary industries are subjected to additional procedures. These industries develop capital equipment for the production of both consumer and non-consumer goods. The industry is further subdivided into the following categories:

- **Large Scale Industry** – It necessitates considerable capital investment in plants and equipment, provides a varied variety of industries, such as other industrial sectors, has a complex industrial organisation that typically employs qualified experts, and generates a large volume of output. Examples include steel and iron production, petroleum refining, and other sectors.
- **Small Industry** – It is characterised by the non-durability of industrial goods and cheap cost of capital in plants and machinery, which may include nonstandard things like customised or design work [1]. Manufacturing of plastics, textiles, food processing, and other sectors are examples.

1.1.3 The Tertiary Sector

The tertiary industrial sector, sometimes known as the service industry, combines industries that provide services or generate revenue without producing tangible items. This sector is often a mix of government and private firms in free market and mixed economies.

Real estate services, banking and finance services, communication services, and so on are examples. An old, well-known anecdote on industrial control.

Throughout history, individuals have gathered to discuss and learn about strategies to exercise organisational control to claim moral authority and political power. Norms concerning effective conduct patterns have arisen through such exchanges and some of these norms have been codified in codes, principles, laws, adages, edicts, and maxims, or the discursive artefacts that people employ to exert political power and claim moral authority.

People have created stories tying situational facts to behavioural norms and institutional settings in order to explain how an organisation exercised control at a specific time and location. People's assumptions about organisational control have evolved throughout time and these assumptions have shaped the stories they have told. This examines how organisational control has been viewed throughout history.

We will start with one of the few times that contemporary organisational scholars have directly addressed the content of this vast human legacy. Rindova and Starbuck (1997) go into detail about how the ancient Chinese viewed organisations and how they employed agency relationship conceptualizations to build different strategies to wield organisational control. Then, we will look at how the exercise of organisational control evolved over time in eighteenth-century Europe and then America [2]. Specifically, we show how industrial bureaucracies arose and how people often resisted the organisational limitations that came with them. We look at how this resistance sparked efforts to make organisational governance more democratic and sensitive to human needs.

1.2 Historical Perspective in Terms of Control

The factory was the first expression of the capitalist firm, which was created by capitalism. Historical events have indicated that important alterations in business organisation have coincides with industrial revolutions since their inception. The British Industrial Revolution (BIR) gave birth to the factory and the Second Industrial Revolution (SIR) gave birth to the massive contemporary business firm by the 1920s.

A multidivisional kind of business organisation is the name given to this type of firm structure (M-form). Another key transformation in business organisation is occurring as a result of today's ICT revolution: "huge vertically integrated" firms are becoming flatter, decentralised, and organised in "semi-autonomous project-based teams". This revolutionary business organisation is commonly referred to as a project-based firm revolution in the literature [3–7].

What is driving these huge organisational changes in businesses? Do these changes put the firm's core values in jeopardy? These are clearly questions about the evolution of a firm and they are questions that are not addressed in post-coasean conceptions of the firm (e.g., Williamson, 1985 [13]). Nonetheless, there is little doubt that studying a firm's growth can aid in gaining a deeper knowledge of it.

In the context of the co-evolution of social and physical technologies, a historical study of the firm's evolution will be developed [14, 15].

The argument will be that the history of the company must be regarded as part of this co-evolutionary process, which has been mostly driven by big changes in physical technology, or macro-inventions in Mokyr (1990) [16] terms, as historical events reveal. The above three firm organisations must be regarded as mutants according to the theory of the firm and the fundamental difference between them is connected to the shift in the mix of low and high-powered incentives (Williamson, 1985) applied within a given mutant-firm.

Human creation procedure also converts as an output of scientific enhancement. The industrial insurrection refers to a shift in production technology which is vastly unlike its old cohort. People's works circumstances and lives were dramatically altered by new industrial technologies. What was the industrial insurrection like and where do we stand? "From the First Industrial Revolution to Industry 4.0" is the title of a new book.

1.2.1 First Industrial Revolution

The utilisation in the eighteenth century of mechanisation and steam power in production ushered in the 1st Industrial Revolution [3]. The automated version, which previously employed simple revolving wheels to generate threads, now produces 8 times the total in the same amount of time. Steam's power was also well acknowledged. The harnessing of human output for industrial goals was the most significant step forward in increasing human productivity. The steam engine could be utilised to power weaving looms instead of muscle power. Next, an enormous swap occurred as an output of enhancement such as the steamship and the steam-powered locomotive that allows passengers and things to move a great displacement in minimum time.

1.2.2 Second Industrial Revolution

The search for electricity and congregation line production announced the 2nd Industrial insurrection in the 19th century. Henry Ford took these concepts and applied them to the production of automobiles, remarkably changing the factories. Initially, a complete automobile was developing at only a station, but now automobiles were produced in parts on a conveyor belt, which is rapid and more economic.

The Second Industrial Revolution began in the nineteenth century with the creation of electricity and the introduction of construction lines.

At a butchery in Chicago where pigs were stretched on conveyor belts and every butcher only accomplished half of the task, Henry Ford (1863-1947) developed the concept for mass production. Henry Ford took these ideas and applied them to the automobile industry, drastically altering it in the process. Previously, a whole vehicle was built at a single station, but now, cars are built in stages on a conveyor belt, which is significantly accelerated and less expensive.

1.2.3 Third Industrial Revolution

The foundation of "memory-programmable controllers and computers to partially robotize activity kicked off the Third Industrial Revolution in the 1970s of the twentieth century". We may now robotize a whole industrial activity without the requirement for personal involvement in regards to the adaptation of these modifications. This is exemplified by robots that follow pre-defined commands without requiring personal intervention. In the 1970s, the use of memory-programmable controllers and computers ushered in the Third Industrial Revolution. In regards to technological advancements, we can now automate an entire production process without the requirement for human participation.

1.2.4 Fourth Industrial Revolution

During the 1st Industrial Revolt, water and steam were used to mechanise industries and thee Secondary used electric power to accomplish mass production. The Third used electronics and computer technology to automate production. A Fourth Industrial Revolution is presently being developed upon the Third Industrial Revolution, the digital revolution. It is explained by a merging of technology that makes the boundaries between the biological, digital, and physical realms.

Currently, the 4th Industrial Revolution is underway. The application of data and communication technologies in the manufacturing business is referred to as "Business 4.0." It is built on the accomplishments of the 3rd Industrial Revolt. A network link powered computer-based manufacturing system, which has a digital counterpart on the Internet in certain aspects can communicate with another system and get data about each other as a result of this. This is a step forward in the process of manufacturing robotization.

All systems are linked, resulting in "cyber-physical production systems" and sophisticated businesses including creation systems, goods, and people. All systems are linked and the outcomes in "cyber-physical production systems" and smart industries, in which manufacturing equipment, and people interact in a network and production is almost independent.

When these authorizations are combined, "Industry 4.0" has the ability to significantly enhance manufacturing settings. Self-organizing logistics include machines that can discover issues and inform maintenance operations on their own or self-organizing management that adapts to unplanned moves in production.

It also has the capability to influence people's work routine. People may be drawn to smarter networks as an outcome of Industry 4.0 that might lead to more efficient work.

The 4th Industrial Revolt, like its predecessors, has the ability to raise world income levels and modify people's lives all throughout the globe. Consumers who can bear and get permission to the digital world have reaped the biggest benefits thus far, as technology has allowed new items and services that increase the enjoyment and efficiency of our lives. Any of these operations, such as hailing a cab, booking a ticket, buying a product, watching a movie, or playing a game, may now be completed remotely. Technological developments will provide a supply-side miracle in the future, resulting in long-term increases in efficiency and production. Communication and transportation and costs will fall, while arrangement and world supply chains will enhance and trade costs will fall, opening up new markets and propelling economic development.

1.3 Future of Industry

1.3.1 Edge Computing

The usage and collection of data distinguishes the third and fourth industrial revolutions. We need technology that allows us to use production information more quickly at any time, regardless of location, in an increasingly competitive and international environment.

Edge computing facilitates smooth management of massive amounts of data by allowing information to be processed at the network's edge, near to the machine that generates the data, lowering processing latency, for example in [6, 7].

Furthermore, because this data has already been processed, the data centre receives lesser volumes, resulting in bandwidth savings.

1.3.2 Additive Manufacturing

3D printers that use new materials will be able to create the unthinkable and will transform industrial processes and systems. This will result in significant cost savings when creating prototypes or small batches [9].

It is now possible to print any element in a variety of shapes and sizes in a completely personalised manner using a variety of materials, as well as to make models and functional parts in small series at extremely low rates, thanks to 3D printing.

1.3.3 5G

5G can improve automation and information transfer speed in the manufacturing process. This will improve the performance of autonomous machines by utilising 5G's "Network Slicing" capability, which allows the network to be divided into subnets.

Network Slicing will tailor connectivity to individual requirements and circumstances. More network resources will be allocated to the machines thanks to the combination of 5G and industrial automation. The full potential of 5G will then be available to the industry's automated sector [11, 12].

Wireless communication speeds are 10 times quicker than earlier creations, rivalling fiber-optic cable speeds, and 5G is the basic mobile technology able to provide broadband wireless technology to attached devices. By 2025, it is predicted that there will be 25 billion linked Internet of Things (IOT) devices. The competition to make 5G commercially available in every country is heating up. When it occurs, the way we communicate—and live—will shift dramatically.

IOT, driverless vehicles, perverted reality, smart towns, mission-complicated manufacturing, 3D videos, remote healthcare, and regenerative medicine are all expected to benefit from 5G [11]. These and other 5G-enabled applications have the potential to revolutionise the legal, risk, and regulatory landscapes, ushering in the Fourth Industrial Revolution as a whole. The Fourth Industrial Revolution will combine sophisticated technologies with high-speed wireless communication to blur the barriers between the biological, digital, and physical, domains. Previous industrial revolutions harnessed the power of steam, electric, and information technology. The ramifications for our systems will change how we create, manage, and govern our planet and ourselves.

1.3.4 Artificial Intelligence

The framework of Industry 4.0 cannot be restricted solely to the abovementioned new technologies.

Artificial intelligence (AI) is at the heart of the factory's internal transformation. AI enables machines and robots to interact with one another and, more importantly, to learn from the numerous situations that arise during the manufacturing processes, resulting in higher productivity and lower costs for enterprises.

However, in the automotive business, artificial intelligence will result in the exponential expansion of driverless systems, which would have a detrimental impact on the industry [10].

As the hype surrounding AI has risen, dealers have been attempting to demonstrate how AI is applied in their goods and services. For example, machine learning is usually referred to as AI. AI demands the use of specialised software and hardware for building and training machine learning algorithms. Although there is no one programming language that is equivalent, along with AI, Python and Java are two common alternatives [7].

Generally, AI operates by absorbing massive groups of labelled training data, analysing the data for relations and designs, and then using these designs to anticipate further states. A chatbot given text chat samples may learn to have lifelike conversations with persons by evaluating millions of cases, while an image recognition computer may learn to identify and describe things in images by examining millions of instances [9].

Self-correction, reasoning, and learning, are the 3 cognitive activities that AI programming targets.

Many academics began to worry that the symbolic method would be able to replicate all of human cognition's operations, including vision, robotics, learning, and pattern recognition. Several academics began investigating "sub-symbolic" methods to certain AI challenges [3]. Rodney Brooks, a robotics researcher, eschewed symbolic AI in favour of targeting on the basic technical issues that allow robots to survive and learn about their atmosphere. In the mid-1980s, Geoffrey Hinton, David Rumelhart, and others resurrected interest in neural networks and "connectionism." Fuzzy systems, Grey system, Neural networks, theory, revolt computation, and many more methods derived from mathematical or statistics optimization were created in the 1980s.

In the late 1990s and early 2000s, AI progressively regained its reputation by identifying and solving particular challenges. Researchers were able to create reliable results, use more mathematical approaches, and work with people from different professions because of the focused emphasis [4]. AI researchers' solutions were extensively employed by 2000, despite the fact that they were rarely referred to as "artificial intelligence" in the 1990s.

Advances in machine learning and perception were aided by algorithmic, faster computer improvements, and availability to vast quantities of data. Around 2012, data-hungry extreme learning algorithms began to influence accuracy standards. According to Bloomberg's Jack Clark, 2015 marked a watershed move for artificial intelligence, with the number of software projects using AI at Google increasing from "sporadic usage" in 2012 to over 2,700. This, he says, is attributable to a growth in inexpensive neural networks, as well as a rise in research tools and datasets, as well as a rise in cloud computing infrastructure. According to a 2017 poll, one out of every five businesses have "integrated AI in certain services or operations." Between 2015 and 2019, the volume of AI research was enhanced by 50%.

Many academics began to worry that AI was no longer following its original objective of producing adaptable, fully sentient computers. Statistical AI, which is overwhelmingly used to tackle specific issues and includes extremely successful approaches like deep learning, is at the centre of much contemporary research. This worry gave rise to the artificial general intelligence (or "AGI") discipline, which by 2010 included numerous well-funded organisations.

The processes of learning element of AI programming is connected with collecting data and creating rules for moving it to benefited knowledge. Algorithms are methods that provide computer equipment with one-by-one instructions for finalization of a certain work.

1.3.5 Cybersecurity

Let's not forget that cybersecurity entails safeguarding data and systems against serious cyber threats including cyber terrorism and cyber espionage [8].

Maintaining cyber security strategy and operations in the face of ever-evolving cyber threats is a huge issue for private enterprises.

As more business activities are automated and a growing number of computers are utilised to store sensitive information, the need for secure computer systems becomes increasingly apparent. This requirement becomes more apparent when systems and applications become widespread and accessible via an unsecured network, such as the Internet. The Internet is used by governments, corporations, financial institutions, and millions of everyday customers. Computer networks support a vast range of functions that, if disrupted, can be disastrous and could productively destroy these businesses. As a outcome, cybersecurity challenges have evolved into national security issues. Keeping the Internet safe is a difficult task. Cybersecurity cannot be obtained via random seat-of-the-pants tactics; it requires deliberate development [8]. Using software engineering ability to tackle the issue is a practical start. Software developers, designers, developers, and administrators, on the other hand, must be aware of the risks and security problems that come with creating, developing, and implementing network-based software.

References

1. Vaidya, S., Ambad,P. Bhosle, S.,Industry 4.0 – A Glimpse, Procedia Manufacturing, 20, 233–238, 2018.
2. Rüßmann, M., Lorenz, M., Gerbert, P., Waldner, M., Industry 4.0: The Future of Productivity and Growth in Manufacturing Industries, Boston Consulting Group, 1-14, 2015.
3. Field C., Internet-Future Communication within the Industry, IEEE Transactions on Industry Applications, 33(2), 397-401, 1997.
4. Mehta, B. S., Awasthi, I C, Industry 4.0 and Future of Work in India, FIIB Business Review, 8(1) 9–16, 2019.
5. Peneder M., Industry Classification: Aim, Scope and Techniques, Journal of Industry, Competition and Trade, 3:1/2, 109-129, 2003.
6. Sharifkhani, A. Simutin M., Feedback loops in industry trade networks and the term structure of momentum profits, Journal of Financial Economics, https://doi.org/10.1016/j.jfineco.2021.04.028.
7. Yu, Y, Zhang J. Z., Cao Y., Kazancoglu Y., Intelligent transformation of the manufacturing industry for Industry 4.0:Seizing financial benefits from supply chain relationship capital through enterprise green management, Technological Forecasting & Social Change 172,120999,1-14, 2021
8. Randall R.G., Allen, S., Cybersecurity professionals information sharing sources and networks in the U.S. electrical power industry, international journal of critical infrastructure protection 34, 100454, 1-10, 2021.
9. Sisodia, A., Jindal R., A meta-analysis of industry 4.0 design principles applied in the health sector, Engineering Applications of Artificial Intelligence 104, 104377, 1-15, 2021.

10. A. Sircar, K. Yadav, K. Rayavarapu *et al.* Application of machine learning and artificial intelligence in oil and gas industry, Petroleum Research, 1-13, 2021, https://doi.org/10.1016/j.ptlrs.2021.05.009.
11. Guo,Q., Xu Q., The economic benefits of agglomeration of the internet of things industry based on 5 G network and markov chain, Microprocessors and Microsystems, 1-5, 2020, https://doi.org/10.1016/j.micpro.2020.103438.
12. Grillitsch, M., Hansen T., Green industry development in different types of regions, EUROPEAN PLANNING STUDIES, 27(11), 2163–2183, 2019.
13. Williamson, O. E. (1985), The Economic Institutions of Capitalism, Free Press, New York.
14. Nelson, R. R., SAMPAT, B. H. (2001), "Making Sense of Institutions as Factor Shaping Economic Performace", Journal of Economic Behavior and Organization, 44, pp. 31-54.
15. Nelson, R. R. (2002), "Bringing Institutions into Evolutionary Growth Theory", Journal of Evolutionary Economics, 12, pp. 17-28.
16. Mokyr, J. (1990), The Lever of Riches: Technological Creativity and Economic Progress, Oxford University Press, New York, London.

2

Industrial Boiler Safety Monitoring System

G. Boopathi Raja

Department of Electronics and Communication Engineering, Velalar College of Engineering and Technology, Erode, Tamil Nadu, India

Abstract

Nowadays, most accidents in industries happen due to poor boiler maintenance. The major boiler accidents that occur in industries are due to several factors. Some of them are water dry-up or poor maintenance, as well as due to high pressure and temperature. There are several existing systems and procedures available, but each technique has its limitation. The drawbacks faced by the existing systems are increased maintenance cost, lack of accuracy, no automation, etc. Many studies and research have been performed to overcome this problem. In this chapter, a new methodology IoT-based Industrial Boiler Safety Monitoring System has been proposed. The important motivation of this proposed framework is to continue monitoring the temperature, pressure, CO level, and toxic content level continuously in the boiler system. The measured values are to be transmitted to the control room through Cloud-based services and/or wireless devices. Based on the obtained data, a decision may be made by the control room. If the level exceeds the threshold, it triggers the entire unit to shut down. Also, this proposed system helps to save energy in industries by using a co-generation system. The proposed system uses a microcontroller and visual application for controlling and remote monitoring.

***Keywords*:** Boiler monitoring, CO level, industrial boiler safety, thermocouple, toxic content

Email: g.boopathiraja14@gmail.com

Vipin Chandra Pal, Suman Lata Tripathi, and Souvik Ganguli (eds.) Industrial Control Systems, (15–36) © 2024 Scrivener Publishing LLC

2.1 Introduction

Industrial accidents frequently happened because of exploding boilers. It can affect industrial production, cause heavy damage to surroundings, kill people and cause severe injuries, and damage the plant. These exploding accidents resulted because of several technical and monitoring issues. These issues include inadequately planned boilers, wrongly sited, or wrongly installed boilers. Unfortunately, the accidents may also occur due to boilers that may not correctly function or are poorly maintained [1, 2].

The chief of a working environment must deal with the risks related to that working environment. If there is a boiler establishment on his/her premises, he/she should guarantee it stays safe. The basic requirement of an efficient and reliable boiler needed for some organizations is to focus more on the downtime caused by a failed boiler. This downtime may cause a major problem in the entire functionality of the industry on production, plant, and individuals.

The word 'boiler' corresponds to a variety of equipment. It includes boilers ranging from basic household hot water boilers to advanced boilers used in industries. The role of boilers used in industries or any other power production plant is to convert fossil fuel into electric power. Mostly, the boilers used for household purposes or domestic boilers are unable to produce steam. They must operate at reasonably lower pressure. Instead, few of the boilers presently work at the higher pressure of the approaching cold-water mains. They may still be far beneath the typical working pressure of steam-raising boilers [3]. This flyer applies to steam and high-temperature heated water kettle plants. The industrial boiler can work with a temperature of 110 °C and above 0.5 bar pressure. It depicts a cycle and as a supervisor, he/she should survey whether the boiler establishment can keep on being worked safely.

For steam and hot water boilers, generators, pumps, burners, and other steam plant equipment, remote access monitoring offers actionable performance data. Alerts can notify users of unanticipated changes in boiler temperature or pressure, fuel level, or other operational issues, reducing reaction times considerably. Remote boiler monitoring can also give useful performance data that can be utilized to optimize boiler efficiency and save fuel expenditures.

Integration of the Internet of Things (IoT), Artificial Intelligence (AI), and blockchain technology play a major role in Industry 4.0 [4, 5]. For enhancing the new industrial revolution, it is necessary to update the features of boilers used in the present scenario.

The overview of the chapter is described as follows. Section 2.1 briefly explains the introduction of the boiler. Section 2.2 demonstrates the factors affecting the performance of the boiler and also the classification of boilers. Section 2.3 describes the proposed methodology to monitor the performance of the boiler. Section 2.4 discusses the various hardware modules used in the proposed methodology. Section 2.5 explains the concluding points and also the future scope of the boiler monitoring system is described.

2.2 Boiler Definition

The steam boiler is a closed mechanical vessel and it is an essential module in most industries. This is utilized in industries where water is boiled and vaporized. It changes over into steam at a pressure extremely higher than nominal atmospheric pressure. A boiler is the greatest and most basic aspect of any conventional industrial unit, especially a thermal power plant. This is also called a steam generator [6]. As per the IBR Act 1923, the Independent organization Indian Boiler Regulation proposed the meaning of a boiler. It states that a steam boiler implies any closed vessel exceeding 22.75 liters in limit which is utilized expressively for producing steam under pressure.

2.2.1 Steam-Raising Plant and Boilers

This type of boiler is widely used in practice in most industries, factories, and several production plants. It is also called a shell and cylinder boiler. The extremely hot gases from the burner pass to the furnace tube. Afterward, it passes back through the smoke tubes and finally, it reaches the flue. The water is boiled by the cylinders and it is done by the primary pressure containing part of the boiler called the shell. The regular vertical boiler is found in various manufacturing plants for either creating extremely heated water or steam.

2.2.2 Factors Affecting Safe Operation of Boilers

All the above-listed factors in the Figure 2.1 are important and must be addressed to ensure the safe functioning of a boiler. On the off chance that at least one of these quits working or is lacking or missing, this could lead to a breakdown of the boiler or catastrophic collapse, making wounds and huge damage to the property [7]. It is decided whether the dangers from

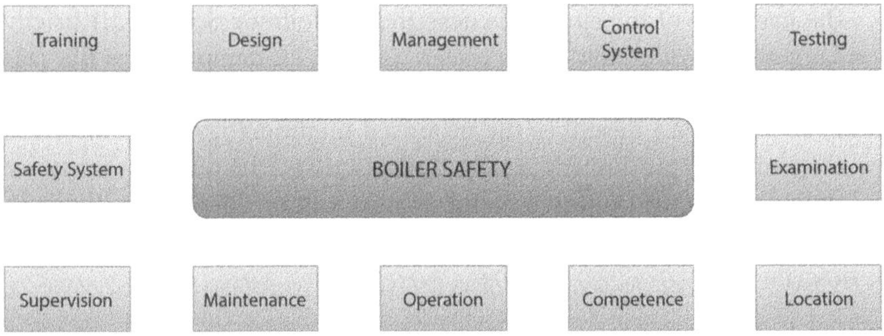

Figure 2.1 Factors affecting safe operations of boilers.

the boilers are at satisfactory levels or if any extra measures are needed to accomplish this. This will require an orderly evaluation of any dangers that might be available and the control gauges set up to address them [8].

A risk assessment does not have to incorporate unnecessary detail and in any case, the discoveries and any action taken ought to be recorded. The specific considerations will rely upon the site, yet may incorporate the following [9]:

a. The boiler detail, condition, and its related plant
b. The way in which the boiler is worked and maintained
c. Any safe frameworks of work, including management and supervision
d. Reliable electronic control framework

2.2.3 Need for Boiler

The most widely recognized capacity for any boiler does not depend on whether it is a private, residential, or industrial boiler. The basic requirement of a boiler is to fulfill essential heating mechanisms for residential purposes, corporate offices, medical clinics, business complexes, and so forth. The boilers works with similar essential capacities and components that cooperate to make a contained, heat-producing burning cycle [10, 11].

Boilers use flammable gas from gas lines going through the corresponding path. It utilizes this gas to fuel the ignition cycle, creating enormous heat and appropriation all through a structure. The boiler framework depends on a burner to start the ignition cycle and afterward, heat as steam or high temperature water travels through the framework utilizing siphons, radiators, and warmth exchangers [12, 13]. Boiler manufacturers are utilizing quickly improving innovations to manufacture gear that is cost-effective, naturally amicable, and ground-breaking [14].

2.2.4 Applications of Boiler

1. Operating steam engines
2. Industrial measure work in synthetic designing and chemical engineering
3. For delivering high temperature water needed to be provided to rooms in freezing zones
4. In warm force stations
5. Operating steam turbines
6. Operating responding pumps
7. The heat substance of the steam is enormous and subsequently, it is reasonable to measure warming in numerous businesses like sugar factories, material plants, dairy industry, and in concoction ventures, [15, 16]

2.2.5 Useful Terms

The following terms are useful for a better understanding of the construction and working principles of the boiler.

a. Boiler Shell
It is the outermost cover of the boiler. It is made up of a cylindrical hollow-shaped body with a number of metal (or alloy) plates bolted or welded together.
b. Furnace (Heater)
The important aspect of the furnace present in the boiler is that the heat is produced when the fuel is burned. This extreme heat is used for the enormous steam generation in the boiler. The furnace is also known as a heater.
c. Grate (Mesh)
The grate is sometimes called Mesh. It is a portion of the boiler where the fuel is allowed to burn. This comprises of a mix of a few cast-iron bars. It is placed in the form that the fuel might befall on it. Space must be kept constantly and given in the middle of two continuous bars so that it may allow streaming into fuel from above the grate. Finally, the ash pit placed under the grate is allowed to collect ashes. It might be square, rectangular, or circular to fit.

d. Grate Area (Mesh Territory)

The territory of the grate whereupon the fuel is consumed is called the grate zone. It is otherwise known as the grate area or mesh territory. The grate zone is constantly estimated in square meters.

e. Heating Surface

The heating surface is nothing but the anterior or outside portion of a steam boiler. It is placed properly where hot gases are allowed to expose on one side and water on another side.

f. Water Space

The portion or area of the boiler or evaporator where the water is occupied, known as water space.

g. Steam Space

The unoccupied portion of the boiler is called steam space. This empty space is required for handling the generation of steam in the evaporator until it is carried out in the steam pipeline.

h. Flue Gases

These gases are extremely hot and created because of the mixture of gases in the heater or furnace. These mixtures normally contain Carbon dioxide (CO2), Carbon monoxide (CO), Nitrogen (N2), and water fumes (H2O). These gases incorporate total and fragmented results of the ignition of fuels. Flue gases are also known as vent gases or hot gases.

2.3 Classification of Boiler

Several boiler designs are used in practice to full fill the requirements of customers and meet the specifications of industries. However, they might be grouped by the accompanying ways:

a. According to the number of tubes present in the boiler:

i. Multi-tube Boiler
   The Cochran boilers belong to the category of multi-tube boilers. These boilers might be named multi-tube heaters since they have various pipe tubes.
ii. Single Tube Boiler
   A Cornish boiler is the best example for single tube boiler. It might be named as a solitary tumbler boiler since it has just a single pipe tube.

b. Based on the utilization, boilers are categorized into:

i. Locomotive Boilers
   Boilers designed specifically and utilized in most steam engines belong to this category.
ii. Stationary Boilers
   The stationary boilers are used for the generation of thermal power. Also, these boilers are preferred for process work such as that in the leather, sugar, textile, and chemical industries.
iii. Marine Boilers
   Steamships use a few boilers to fulfill the requirements falling into this category.

c. As per the nature of the fuel, boilers are categorized as follows:

1. Fuel-fired boilers
2. Gas-fired boilers
3. Liquid fuel-fired boilers
4. Electrically fired boilers
5. Nuclear-fired boilers

The Wilcox and Babcock boilers require a strong or vaporous fuel. Volex boilers need oil as fuel.

d. Based on the nature of pressure, boilers are named:

i. Low-Pressure Boiler
   These boilers generate steam in the range of 15 to 20 bar pressure. This is utilized to measure warming.
ii. Medium Pressure Boiler
   The medium pressure boiler produces steam with medium pressure ranging from 20 bar to 80 bar. This type of boiler may be utilized in the process of heat or power generation.
iii. High-Pressure Boiler
   A high pressure boiler utilizes extremely high pressure above 80 bar.

e. ***Based on the axis of the boiler shell***
Boilers are classified into two types based on the situation of the axis of the boiler shell. They are:

i. Horizontal Boiler
The horizontal boiler is the one in which it has a horizontal boiler axis. The Lancashire boiler belongs to the category of horizontal boiler.

ii. Vertical Boiler
The vertical boiler is the one which has a vertical boiler axis. The Cochran boiler is the best example of a vertical boiler.

f. ***As per the flow of gases, the boilers are categorized as follows:***

i. Fire-tube Boiler
The boilers wherein hot gases are delivered by a blend of fuel in the boiler furnace are called Fire-tube boilers. Based on this procedure, the hot gases follow the chimney and pass through various cylinders such as fuel cylinders that are immersed in water. The hot gasses are transferred to the water through the chambers of cylinders, hence excess heat in the gases leads to the generation of steam. The Locomotive boiler and Cochran boiler are suitable for this category. A fire-tube boiler can also be called a smoke tube boiler.

ii. Water-tube Boiler
In certain boilers, water passes through various cylinders, also called water tubes. Such boilers are known as water-tube boilers. The hot gases generated due to the fuel ignition in the boiler are permitted to pass through the chimney encompassing the cylinders. The hot gases provide enormous heats that are allowed to pass to the water through the dividers or chambers of the water tubes. The Wilcox evaporator, Bab cock, and Benson kettle fall under this category.

g. ***Based on the circulation of water, boilers are classified under two categories:***

i. Free Circulation Boiler

Normally in all boiling vessels, the heat is transferred starting with one spot, then onto the next. This is possible through convection since water is a poor conductor of heat.

Suppose if the vessel containing water may be boiled at its base. The water available in the base portion is heated quickly, hence its density gets diminished. On other hand, the density of water in the upper portion of the vessel is greater. Accordingly, the less dense water at the base part of the vessel moves upward. Also, nearly more dense and cold water at the upper portion of the vessel comes down to reach the base spot. Subsequently, convection current is generated in the water until the difference in temperature in the entire water level turns into the equivalent.

The technique for the circulation of water depicted above is known as free circulation. The boilers such as the Babcock boiler, Lancashire boiler, Wilcox boiler, and so on follow this free circulation approach.

The benefits of free circulation approaches are:

1. The free circulation of water assists with keeping up a uniform temperature true wherever inside the boiler so that the inconsistent extension of different parts of the boiler is forestalled.
2. The free circulation of the water provides steam from the heating surface of the boiler when it is framed. On the off chance that steam does not get away rapidly after its arrangement, the standards do not remain continually in contact with water and accordingly, these plates might be overheated.

ii. Forced Circulation Boiler

Forced circulation is otherwise called constrained circulation. In constrained circulation, the pumps are utilized to keep up the nonstop progression of water in the boiler. In these cases, the flow of water happens

because of pressure made by the pump. This constrained or forced circulation framework is embraced in more high pressure, high limit boilers, which are all water tube type boilers.
The benefits of constrained circulation are:

1. It provides the maximum rate of heat transfer from the extremely hot gases into water.
2. Tubes are kept at a reasonably small width. This diminishes the general load of the boiler.
3. The requirement for boiler drums will be gradually reduced.
4. Lower scale arrangement in the boilers is required.
5. 'The required steam can be immediately produced.
6. The load fluctuation must effectively meet without getting the assistance of any advanced or complex gadgets.
7. There may be a lesser chance of overheating of the boilerplates.
8. There is a less nominal weight per unit mass of steam, which is always generated.

2.3.1 Types of Fuels Used in Boilers

The fuels used in boilers are categorized as solid, liquid, or gas.

a) Solid Fuels
There are several resources available for solid fuels. These are cheaper compared to liquid and gaseous types. Some of them are listed as follows: Coal, Wood, Rice Husk, Pet coke, and Briquettes (a block of compressed coal dust).
b) Liquid Fuels
An example of liquid fuels is furnace oil and LDO (Light Diesel Oil).
c) Gaseous Fuels
These fuels can be utilized to carry out the combustion for any specific application. The gaseous form of fuels includes the following categories:

i. LNG (Liquefied Natural Gas)
ii. LPG (Liquefied Petroleum Gas)
iii. PNG (Piped Natural Gas)

Figure 2.2 Outline of proposed framework.

2.4 Proposed System

The main objective of the proposed framework is to continuously monitor the temperature, pressure, water, and toxic levels present in the boiler system.

The basic block diagram of the proposed framework Low-cost Industrial Boiler Safety monitoring system was shown in Figure 2.2. The proposed framework consists of two modules, they are:

1. Transmitter section
2. Receiver section

The transmitter section is attached to the boiler, whereas the receiver section is placed in the control room. The transmitter section consists of several sensors such as a temperature sensor, pressure sensor, thermocouple, water level indicator, and gas detector to monitor the toxic content present in the emitted gases. These measured values are to be transmitted to the receiver section, that is, the control room through wireless devices. The wireless devices should be Bluetooth, ZigBee, RF transceiver, etc. The choice of the transceiver is decided by the distance between transmitter and receiver and also the type of application where the boiler is used. In the receiver section, further decisions need to be made based on the received parameters.

2.4.1 Transmitter Section

Figure 2.3 shows the block diagram of the transmitter module placed along with the boiler. It consists of a microcontroller, thermocouple, temperature sensor, pressure sensor, toxic gas detector, water level indicator, RF

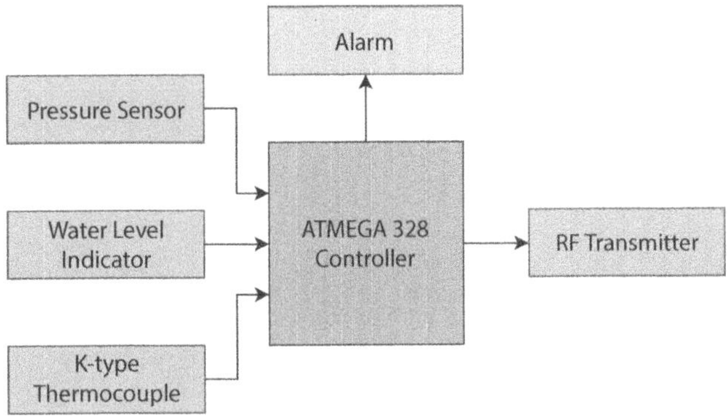

Figure 2.3 Transmitter section.

transmitter, and alarm circuit. These sensors continuously capture information such as temperature and pressure of the boiler, steam flow, toxic content present in the steam, and water level in the boiler. The sensed analog data is processed in the microcontroller. The microcontroller provides the information to the transmitter section in digital form. The microcontroller may be Raspberry pi, Arduino UNO, 8051 microcontrollers, ATMega microcontroller, or PIC microcontroller. The choice depends on the type of application where the boiler is used, the capacity of the boiler, the number of sensors to be interfaced, etc. In this application, an ATMega 328 microcontroller is preferred because it has attractive features at a low cost and provides better performance.

The digital information received from the transmitter section may either be stored in Cloud storage for future use or transmitted directly to the receiver for immediate action.

2.4.2 Receiver Section

Figure 2.4 shows the block diagram of the receiver module located in the control room, which is placed far away, which is a reasonable distance from the transmitter section [20–22]. It consists of a Microcontroller, signal conditioning and monitoring unit, DSP processor, storage unit, RF receiver, PC, visual basic application, and alarm circuit. The information received from the boiler is converted into a suitable form and analyzed by using signal conditioning and a monitoring unit. If the received parameter crosses the threshold, then the decision will be made with the help of the ATMega Microcontroller. Also, the parameters are to be continuously updated and displayed on the PC through the visual basic application.

Figure 2.4 Receiver section.

2.5 Hardware Components

2.5.1 ATMEGA 328 Controller

The Atmel pico Power ATmega328/P is a low-power CMOS 8-digit microcontroller. It is dependent on the AVR improved RISC architecture. The ATmega 328/P achieves the desired throughput closer to 1 MIPS per Megahertz frequency by executing more powerful instructions per single clock cycle. This engages the system designer or developer to improve the gadget for power utilization versus operating speed.

2.5.1.1 Features of ATmega 328/P Microcontroller

The ATmega 328/P Microcontroller belongs to the low power, high-performance Atmel AVR 8-bit Microcontroller Family.

Advanced RISC Architecture

a. It handles 131 powerful instructions
b. Most of the instructions are executed only in a Single Clock Cycle
c. It uses 32 x 8 General Purpose Registers
d. It provides up to 20 MIPS Throughput at 20MHz
e. It has a 2-cycle On-chip Multiplier
f. ATmega 328P supports the full static operation

High Endurance Non-Volatile Memory Segments

a. It has 32KBytes of In-System Self-Programmable Flash Program Memory
b. It has a data retention facility of about 20 years at 85°C and/or 100 years at 25°C
c. It supports Write/Erase Cycles with 10,000 Flash/100,000 EEPROM

d. It uses 1KBytes EEPROM and 2KBytes Internal SRAM
e. Programming Lock for Software Security
f. Optional Boot Code Section with Independent Lock Bits

- True Read-While-Write Operation
- In-System Programming by On-chip Boot Program

Atmel QTouch Library Support

a. QTouch and QMatrix Acquisition
b. Up to 64 sense channels
c. Capacitive Touch Buttons, Sliders, and Wheels

Peripheral Features

a. Two 8-bit Timer/Counters with Separate Prescaler and Compare Mode
b. One 16-bit Timer/Counter with Separate Prescaler, Compare Mode, and Capture Mode
c. One Byte-oriented 2-wire Serial Interface (Philips I2C compatible)
d. One Programmable Serial USART
e. Real-Time Counter with Separate Oscillator
f. Six PWM Channels
g. One On-chip Analog Comparator
h. Two Master/Slave SPI Serial Interface
i. Programmable Watchdog Timer with Separate On-chip Oscillator
j. Interrupt and Wake-up on Pin Change
k. 6-channel 10-bit ADC in PDIP Package (Temperature Measurement)
l. 8-channel 10-bit ADC in TQFP and QFN/MLF package (Temperature Measurement)

Special Microcontroller Features

a. External and Internal Interrupt Sources
b. Internal Calibrated Oscillator
c. Power-on Reset and Programmable Brown-out Detection
d. Six Sleep Modes: Idle, ADC Noise Reduction, Power-save, Power-down, Standby, and Extended Standby

I/O and Packages

a. 28-pin PDIP, 32-lead TQFP, 28-pad QFN/MLF and 32-pad QFN/MLF
b. 23 Programmable I/O Lines

Additional Features

a. The range of Operating Voltage is from 1.8V to 5.5V
b. Operates in the temperature varying from - 40°C to 105°C
c. The approximate Power Consumption at an operating frequency of 1MHz is about 1.8V at a nominal room temperature of 25°C
d. The current handling capacity of the controller during Active Mode is around 0.2mA, at Power-down Mode is about 0.1μA, and during Power-save Mode is approximately 0.75μA by including a 32kHz Real-Time Clock
e. Speed Grade of ATMega 328P is available with three options: (0 - 4MHz) at 1.8 - 5.5V, (0 - 10MHz) at 2.7 - 5.5V, and (0 - 20MHz) at 4.5 - 5.5V, respectively

2.5.2 Thermocouples

Thermocouples are a typical kind of temperature sensor. It can be utilized in applications ranging from lab tests to modern industrial applications. In any case, it is additionally used in the situation with numerous types of sensors. It is used to obtain the exact information with thermocouples that can be tested.

2.5.2.1 Construction

The two different conductors are combined to form hot and cold intersections that make a thermocouple (Figure 2.5). Keeping up these two intersections at various temperatures creates an electromotive force (EMF), otherwise called the thermoelectric voltage. EMF is estimated in the range of millivolts (mV). It is the result obtained as the product of a physical phenomenon called a Seeback effect. This effect depicts the voltage that results from the thermocouple materials and the temperature contrast between the hot and cold intersections.

The linear approximation of the Seeback effect is described in Equation 2.1.

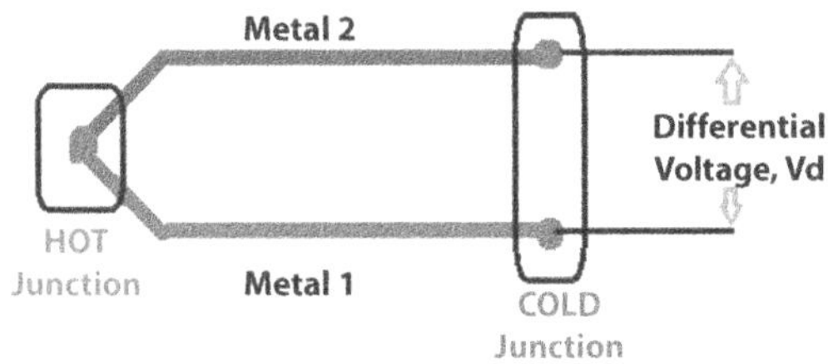

Figure 2.5 Fundamental thermocouple.

$$\nabla V = S.(T_2 - T_1) \tag{2.1}$$

where

∇V – Voltage difference between two non-identical or dissimilar metals

S – Seeback coefficient; it is measured in terms of V/K (commonly in μV/°C)

$(T_2 - T_1)$ – The difference between the temperature at hot and cold junctions

The Seeback coefficient is explicit to the two conductors that are utilized to build the thermocouple. This coefficient is a non-linear function of temperature. The Seeback effect provides reasonable measurement errors by using the linear approximation technique. The recent thermocouple estimation techniques should consider these non-linearities during the construction as well as operation. It is important to note that temperature estimation cannot be resolved exclusively from the EMF created by the thermocouple. Rather, the accompanying three factors related to the thermocouple must be known:

1. Thermoelectric voltage because of the thermal gradient developed among the hot and cold junctions
2. Type of Thermocouples
3. Temperature at the colder junction

If any of these factors are not known, the temperature at the thermocouple sensing junction could not be solved. The updated computation for estimating the temperature at the hot junction of a thermocouple, as shown in equation 2.2, considers the above factors:

$$T_H = \frac{V_{Thermocouple}}{\alpha(Tc)} + Tc \tag{2.2}$$

where T_H – Temperature at the hot junction (in °C)
T_C – Temperature at the cold junction (in °C)
$\alpha(T_C)$ – Seeback coefficient as a non-linear function in terms of T_C (μV/°C)

2.5.2.2 *Factors Impacting Accuracy of Thermocouple Readings*

The following factors may influence the temperature perusing at various phases of estimation and to changing degrees:

a. Thermocouple qualities
b. Colder junction estimation
c. Noisy conditions
d. Linearization

2.5.2.3 *Thermocouple Characteristics*

Now and then, issues with the thermocouple itself lead to erroneous readings. These issues may cause severe damage that happens as a thermocouple age. Some normal issues to know about include:

a. Poor hot junction thermal/electrical connection
b. Series resistance along the lines of the thermocouple
c. Low-grade materials used to develop the thermocouple
d. Electrical separation issues along the length of the leads
e. Thermocouple type

2.5.2.4 *Industrial Thermocouples*

Thermo sensors-based industrial thermocouples are generally utilized in most process-oriented industry applications. These are commonly chosen by deciding based on specific conditions under which it must perform. These conditions have suggested wire and material choices and are assembled in types.

Thermo sensor thermocouple component types consist of the following categories:

a. Type E Chromel-Constantan Thermocouple
b. Type J Iron-Constantan Thermocouple
c. Type K Chromel-Alumel Thermocouple
d. Type N Nicrosil-Nisil Thermocouple
e. Type R Platinum-Platinum 13% Rhodium Thermocouple
f. Type S Platinum-Platinum 10% Rhodium Thermocouple
g. Type B Platinum 6% Rhodium-Platinum 30% Rhodium Thermocouple
h. Type T Copper- Constantan Thermocouple

2.5.3 Pressure Sensors

Real-time sensor data is empowering plants to more likely comprehend their processing cycles and keep them running. Placing the information into the IIoT also allows optimizing how raw materials are requested, handled, and consumed.

Keen assembling is changing how we make, package, and circulate everything, but Industry 4.0 is now revolutionizing how the processing plants work. While the Industrial IoT is subject to availability, generally it is harnessing the raw information and transforming that into operational intelligence, which implies sensors are critical to the whole cycle. While there are numerous sorts of sensors at play here, the pressure sensor is presumably the most assorted and generally sent type of sensor in the IIoT.

2.5.3.1 Monitoring Process Flows

Differential pressure sensors are utilized broadly in process flows where a liquid needs to go through some type of boundary, for example, a channel. Under ordinary conditions, the difference in pressure between the upstream (regularly called the influent or line pressure) and the downstream (emanating) pressure ought to be nil or negligible. As the channel gets hindered with foreign substances, the downstream pressure will diminish, which causes the difference estimated to increment [16, 17].

The output of the sensors can be adjusted to show the most extreme reasonable pressure difference at full scale. For instance, a 4-20mA yield could be adjusted to show 20mA when the pressure contrast arrives at the most extreme reasonable, yet read 4mA when the pressure distinction is nil.

Differential pressure sensors measure the contrast between the effluent (outward) and influent (inner) pressures of a channel, which ought to be nil under typical working conditions, yet will ascend as the channel gets stopped up.

2.5.3.2 Estimating Safe Levels in Liquid Tanks

Submersible pressure sensors are affirmed for use intrinsically in safe areas. It can be utilized to gauge fluid pressure of up to 30 PSI with either a voltage or current output in the range of 4-20mA. Situating a submarine pressure sensor at the lower part of a tank would give an exact value of the substance of the tank, accordingly cautioning laborers or the cycle control framework when the level in the tank falls under a suitable lower limit [18].

The pressure at the lower part of a tank is ordinarily called hydrostatic or head pressure. This pressure was estimated in units of distance such as feet, inches, or meters of water sections. For instance, 27.670 inches of water section is around equivalent to 1 PSI at 100°F. The pressure estimated is subject to the height of the tank, rather than its shape or the volume of the fluid. Thus, it is significant that the sensor is put at the lower part of the tank (rather than mostly down).

Process cycle regulators can ascertain the degree of fluid in a tank by estimating the hydrostatic pressure, which is all the more precisely estimated when the thickness or density of the fluid is likewise known.

2.5.3.3 Managing Control Loops

Just as being utilized to screening measures, pressure sensors are frequently instrumental in the control loop [19]. This is especially applicable in the utilization of hydrodynamics, where pressurized liquids are utilized to apply exertion in presses or lifts for instance.

The sensors are regularly little, especially those dependent on MEMS innovation. They can measure less than 2mm on each side, yet are equipped for estimating absolute pressure in the area of 20 Bar or more. This makes them appropriate in the scope of utilizations, including automotive and clinical.

2.6 Conclusion and Future Scope

The various boiler parameters are continuously monitored and recorded by utilizing various sensors. It has been constructed by integrating features of all the hardware components and software used. Each module in both the transmitter and receiver section has been reasonably picked and constructed carefully, thus contributing to the effective working of the boiler. Also, by using the highly advanced controller and with the help of recent techniques, the framework has been studied. With this framework, it is identified what must be satisfied in the boiler operation. Also, it supports the better controlling action appropriately. This low-cost framework was designed to monitor the performance of the boiler continuously and it provides more accuracy in results. The framework implemented can be suitable for boiler automation. It may also be used in many other applications. Also, the measured values are displayed in LCD.

By implementing this system, observing and controlling several boiler parameters at the transmitter side is possible. The essential parameters such as humidity, temperature, pressure, gas level, and boiler water level are monitored continuously and displayed the results using the web server. The parameters obtained in the boiler are read at the receiver end by using a visual basic application or visualized through Think Speak. In the future, along with application software, Internet of Things (IoT) and Artificial Intelligence play a major role in handling activities performed in Industry 4.0. The IoT provides a higher speed of control with lower bandwidth, less memory requirement, and low power consumption. The implementation of automation in an industry is also based on big data analytics and cloud computing. This can be used for high-five technologies of hardware-implemented plants. If the internet connection is given to the monitoring system, then the displayed result is visualized in Think speak at any place with an internet connection. The parameters can be analyzed and visualized with the uploaded data using a graphical system.

References

1. Ismatkhodzhaev, S., & Kuzishchin, V. (2017). Enhancement of the efficiency of the automatic control system to control the thermal load of steam boilers fired with fuels of several types. Thermal Engineering, 64(5), 387-398. doi: 10.1134/s0040601517050032.
2. Sabanin, V., Starostin, A., Repin, A., & Popov, A. (2017). Study of connected system of automatic control of load and operation efficiency of a steam boiler with extremal controller on a simulation model. Thermal Engineering, 64(2), 151-160. doi: 10.1134/s0040601517020070.
3. Gbasouzor, A. (2015). Design and Characterization of a Model Pilot Multi-tube for the Transfer of Heating the Fire Tube Boilers in Academic

Laboratories. Transactions on Engineering Technologies, 255-288. doi: 10.1007/978-94-017-9804-4_18.

4. Raja G. B. (2021) Impact of Internet of Things, Artificial Intelligence, and Blockchain Technology in Industry 4.0. In: Kumar R., Wang Y., Poongodi T., Imoize A.L. (eds) Internet of Things, Artificial Intelligence and Blockchain Technology. Springer, Cham. https://doi.org/10.1007/978-3-030-74150-1_8.
5. Raja G. B., Impact of IIOT in Future Industries: Opportunities and Challenges. In: Rana A. K., Ayodeji Olalekan Salau, Sharad Sharma, Shubham Tayal, Swati Gupta (Eds.) Internet of Things: Energy, Industry, and Healthcare, CRC, 1st Edition, 2021.
6. Hou, G., Zhao, W., & Li, S. (2014). Research of Power Plant Boiler Control System Based on Compensatory Fuzzy Neural Network. Applied Mechanics and Materials, 614, 203-206. doi: 10.4028/www.scientific.net/amm.614.203.
7. S. Jaisiva,V. Vijayabharathi, P.Vel, S. Saran, Condition and Monitoring Of Boiler Parameters by Cbm Using Raspberry Pi Controller. (2019). International Journal of Engineering and Advanced Technology, 8(6), 732-735. doi: 10.35940/ijeat.f7982.088619.
8. Raja, G. (2021). Fingerprint-based smart medical emergency first aid kit using IoT. Electronic Devices, Circuits, and Systems for Biomedical Applications, 325-348. doi: 10.1016/b978-0-323-85172-5.00015-0.
9. Industrial Parameters Monitoring System based on Embedded Web Server. (2019). International Journal Of Innovative Technology And Exploring Engineering, 8(9), 2450-2453. doi: 10.35940/ijitee.i8935.078919.
10. Percis, D. E., Nalini, D., & Rama, M.T. (2018). Raspberry Pi Based Coal Mine Safety System.
11. Poonam, R., & Sonali, K. (2017). Industrial Monitoring using Raspberry Pi Technology. International Advanced Research Journal in Science, Engineering and Technology, 4, 112-114.
12. Internet of Things: Smart Home Automation System using Raspberry Pi. (2017). International Journal of Science and Research (IJSR), 6(1), 901-905. doi: 10.21275/art20164204.
13. Adeoti, O., Akomolafe, A., & Adebola, F. (2019). Monitoring Process Variability Using Double Moving Average Control Chart. Industrial Engineering & Management Systems, 18(2), 210-221. doi: 10.7232/iems.2019.18.2.210.
14. Lian, G., Niu, Y., Zhang, X., Lu, Y., & Li, H. (2018). Analysis of Causes of Boiler Accidents in Power Plant and Accident Handling Based on Mathematical Statistics. 2018 International Conference on Engineering Simulation and Intelligent Control (ESAIC). doi: 10.1109/esaic.2018.00012.
15. Water Level Monitoring using Blynk Application in IoT. (2019). International Journal of Recent Technology and Engineering, 8(4), 1676-1679. doi: 10.35940/ijrte.c5358.118419.
16. Dhanraj, J., Ramanathan, K., Priyadharsini, S., Jayaraman, P., & Jothi Sankar, M. (2020). Boiler Temperature and Pressure Monitoring System for Thermal

Power Plant through LabVIEW. IOP Conference Series: Materials Science and Engineering, 988, 012079. doi: 10.1088/1757-899x/988/1/012079.

17. Okhaifoh, J., Igbinoba, C., & Eriaganoma, K. (2016). Microcontroller Based Automatic Control For Water Pumping Machine with Water Level Indicators using Ultrasonic Sensor. Nigerian Journal of Technology, 35(3), 579. doi: 10.4314/njt.v35i3.16.
18. Takaghaj, S., Macnab, C., Westwick, D., & Boiko, I. (2014). Neural-Adaptive Control of Waste-to-Energy Steam Generators. IEEE Transactions on Control Systems Technology, 22(5), 1920-1926. doi: 10.1109/tcst.2013.2292818.
19. Damarla, S., & Kundu, M. (2011). Monitoring of Drum-boiler Process Using Statistical Techniques. International Journal of Chemical Engineering and Applications, 173-180. doi: 10.7763/ijcea.2011.v2.97.
20. Zhang, Q., Yi, H., Yu, Z., Gao, J., Wang, X., Lin, H., & Shen, B. (2018). Energy-exergy analysis and energy efficiency improvement of coal-fired industrial boilers based on thermal test data. Applied Thermal Engineering, 144, 614-627. doi: 10.1016/j.applthermaleng.2018.08.069.
21. Proceeding of International Conference on Intelligent Communication, Control and Devices, Springer Science and Business Media LLC, 2017.
22. Dipak K. Sarkar, Steam Generators, Elsevier BV, 2015.

3

Robust Control of Industrial Rotary System

Naiwrita Dey[1]*, Ujjwal Mondal[2] and Anindita Sengupta[3]

[1]*Department of ECE, RCCIIT, Kolkata, India*
[2]*Department of Applied Physics, Calcutta University, Kolkata, India*
[3]*Department of EE, IIEST, Shibpur, India*

Abstract
Many industrial applications involve rotary machinery. Dynamics of such rotary systems are often subjected to position dependent disturbances. Therefore, the ingrained need for periodic disturbance attenuation is the basic motivation of this present work. This book chapter reviews the existing literature in controller design for such systems. A modified robust approach of finite dimensional repetitive controller design and its synthesis is proposed in the present work. A repetitive controller (RC) can follow any frequency periodic command or reject periodic disturbance if the period of the input is exactly equal to the RC loop delay. The practical system often tends towards instability at higher frequencies. Also, small changes in frequency and phase of reference signal, known as periodic uncertainty, causes instability. The proposed controller has a twofold objective, the first one aiming at the precise tracking performance by disturbance rejection and the other one is a robust stabilization of the overall system. A Lyapunov–Krasovskii functional is chosen to obtain the LMI-based sufficient convergence condition. A DC servomotor and rotary roller pump are the two plant models, considered here for the simulation study. Thee processor in a loop simulation is performed using a Raspberry Pi microcomputer for validation of the proposed controller.

Keywords: Periodic disturbance, robustness, lyapunov, repetitive control, processor-in-loop (PIL)

**Corresponding author*: naiwritadey@gmail.com

Vipin Chandra Pal, Suman Lata Tripathi, and Souvik Ganguli (eds.) Industrial Control Systems, (37–58) © 2024 Scrivener Publishing LLC

3.1 Introduction

Disturbance rejection is one of the sole purposes of feedback control system design. Other control problems like vibration suppression and reference command tracking in position control and it is important to reject persistent disturbances [1]. Different rotary industrial applications such as computer disk drives, rotary actuators, automated machineries, pumps, the rotating engine of helicopter compressor, motor gears, etc. involve periodic operations. Such repetitive kinds of activities may include periodic disturbances in the system [2, 3]. It can be observed that periodic disturbances are encountered in a wide area of applications that are involved across the whole engineering application field. Synchronization of rotational speed of the system and such repetitive disturbances may cause vibrations to the system and can result in damage. Therefore, disturbance attenuation has been one of the prime control objectives for decades. Different approaches of disturbance rejection control are reported in the literature. This chapter begins with an overview of observer based design and an internal model principle based design of disturbance rejection controller, respectively, for known and unknown frequency signals. Application of a disturbance observer in order to estimate and compensate the effect of the periodic disturbance has been studied by many researchers. The disturbance observer was for a magnetic hard drive servo system control [4]. This periodic disturbance includes fundamental waves along with harmonics. Zero gain and zero phase lag filter were experimented with for rotary pumps and compressors with a disturbance estimation method independent of input saturation. Up to fifth order harmonics were taken into consideration [5]. Robust design of the disturbance observer is implemented within H-infinity framework to minimize the input disturbance estimation error which is the ultimate objective of the control system [6]. Disturbance rejection of the peristaltic pump was reported in the literature [7]. A repetitive controller belongs to a special class of controller which works great for periodic signals both for tracking of periodic command and rejection of periodic disturbances. Since the repetitive controller does not perform significantly well if minimal frequency variation is in the disturbance frequency [8], to address this problem different robust design approaches such as adaptive repetitive control and high order repetitive control are discussed in the literature [9, 10]. An adaptive repetitive controller is proposed for reference tracking control of servo systems along with compensation of spatial periodic disturbance signals. A piecewise continuous friction model is obtained to validate the controller performance [11]. Often, rotary machines encounter position dependent time varying periodic kinds of disturbances. Uniform time-sampling based rejection of such disturbances is implemented with a spatial repetitive controller [12]. Steinbuch proposed a higher order repetitive controller and a solution was approached towards the performance degradation problem under variation of frequency. Multiple repetitive control loops are applied [13]. A robust repetitive control approach has been studied for the current control technique of a grid connected inverter system. The implementation of an LCL filter was investigated and with the variation of output side inductance of the same the controller, exhibited robust performance with a comparatively lower value of harmonic distortion compared to conventional PI controllers [14]. Application of higher-order repetitive controllers (HORC) was studied for a grid connected inverter system. Inclusion of a stabilizing compensator along with a compensator for zero phase tracking error has improved the robust stability of the system [15]. As a remarkable result of the higher order RCs, it is was observed that the notch of sensitivity function has been widened at the periodic signal frequency with an implementation of 7th order RC [16]. In a state space framework, a modified HORC structure has been defined. The low pass filter cut-off frequency effect was analyzed with robust synthesis of the feedback control gain matrix with a formulation of LMI [17]. Discrete wavelet transform based robust design of a repetitive controller is also reported in the literature [18]. Application of linear matrix inequality (LMI) was reported to obtain controller gain parameters, satisfying sufficient robust condition. Conventional repetitive control structure is extended to a continuous-discrete robust 2D model [19]. It has been studied that often, stability requirement constrains the bandwidth. As an alternative approach, the robust stabilization problem is reformulated as a linear objective minimization problem of the filter parameter. Later, it is subjected to LMI constraints developed from a Lyapunov function for that particular system [20].

The motivation behind this research work is to overcome the relatively poor performance of conventional controller with periodic disturbance compensation. Repetitive controllers (RCs) are able to reject periodic signals. Many works have been reported in the previous section. However, the effectiveness of the repetitive controller is limited by certain factors such as bandwidth of the plant and order of the controller. Therefore, a robust approach is made to design a repetitive control loop for disturbance attenuation in rotary systems.

The technical contribution of this book chapter can be arranged as follows:

i) In the proposed robust design approach towards the repetitive controller, a multiloop repetitive controller is introduced rather than a conventional one. With two loops, a second order finite dimentional repetitive controller is proposed here. Unlike conventional repetitive controllers, the knowledge of a complete disturbance model is not required here.
ii) The detailed robust stability criteria has been derived analytically from a Lyapunov-Krasovskii Functional Approach. Further, it is verified with simulation results for two different test case studies. Significant improvement in the disturbance attenuation performance is shown. The best achievable closed-loop set point tracking performance is also shown.
iii) For experimental validation of the proposed design, a processor-in-loop simulation is carried out using Raspberry PI and a single board computer. This is a proven approach for the feasibility of the proposed method in real time.

The remaining book chapter is given as follows. In Section 3.2, the multiloop repetitive controller design strategy is proposed. Basic problem formulation is described in Section 3.3. Robust stabilization criteria is obtained based upon Lyapunov functional method in Section 3.4. Sections 3.5 and 3.6 present two different plant models and simulation study of them, respectively. The processor-in-loop simulation process is demonstrated for the proposed controller in Section 3.7 and Section 3.8 concludes the chapter.

3.2 Controller Design

The repetitive controller (RC) has a built in feature that can track any periodic reference command or attenuate any periodic disturbance signal precisely, but the period of signal has to be known, which is not always a convenient option. Often, a little variation is observed in the time period of the signal and time delay block in RC loop, which significantly degrades the tracking or disturbance attenuation performance of the controller. To address this problem, sensitivity of the repetitive controller towards input periodic disturbance has to be reduced and that is the main objective here. A novel design methodology of introducing multiple delay loops is to be introduced unlike a single delay loop in conventional RC. Mathematically, it is proven that RC can track or reject of any frequency periodic signal. In literature it is reported that the individual peak gains are very high around the higher frequency range, resulting in instability. A low pass filter is placed in the conventional RC loop and thus it has been modified to a finite dimensional RC (FDRC) loop ensuring enough of a stability margin along with disturbance rejection ability. A robust design approach is concerned about the notable amount of fluctuation between the period within the FDRC loop delay and the disturbance signal period.

3.2.1 Finite Dimentional Robust Repetitive Controller Using a Multiloop Approach

A multiple finite dimensional memory loops works in a way that cumulates errors for more than one previous test and the controller performs in an enhanced way with robust property in terms of period variation. Thus, a finite dimensional robust repetitive control approach (FDRRC) is proposed here. This multi-loop approach is also termed as a higher order RC. Another major point to ponder is the stability; with a higher order controller having more than one FDRC loop, it has a tendency to attain robustness to disturbance attenuation causing instability. Therefore, robust stabilization along with disturbance rejection performance is the control objective here. The robust repetitive controller (RRC) proposed by Stecinbuch is shown in Figure 3.1 [3]. Implementation of multiple memory loops along with the weight factors has modified the dynamic response of the controller between harmonic frequencies. Selection of weight factor was done further by designing an optimization problem and that was solved using linear programming. The weight factors of the RRC loop shown are to be selected in a way to minimize the infinite norm of the sensitivity function [21].

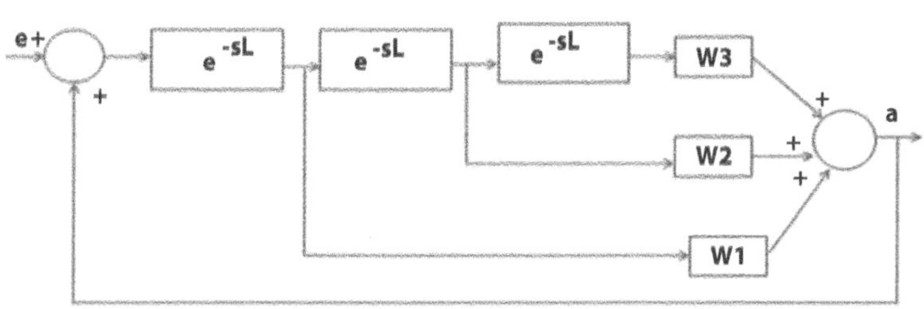

Figure 3.1 Steinbuch's RRC model [11].

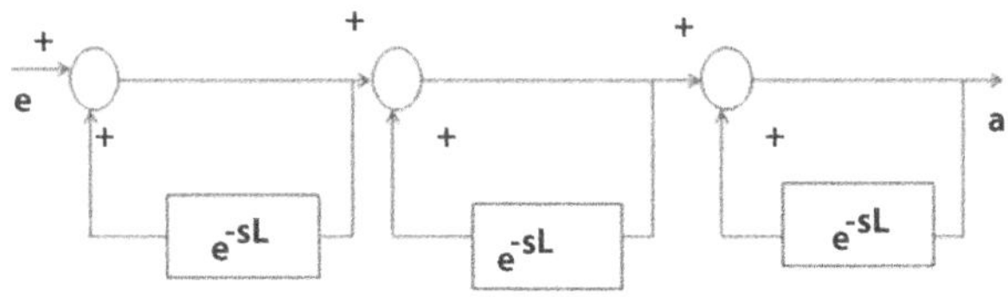

Figure 3.2 Singh's RRC model [20].

Further, this RRC loop has been modified by Singh, as shown in Figure 3.2, where multiple RC loops are cascaded to eliminate the weight factor selection process [22].

Proposed FDRRC Model

A first order low pass filter, $Q_{RC}(s)$, is included so the conventional RC loop results in finite dimensional RC (FDRC). The low pass filter specification is given as follows. The cut off frequency is ω_{RC}.

$$Q_{RC}(s) = \frac{\omega_{RC}}{s+\omega_{RC}}, \quad \left|Q(j\omega)\right| \approx 1, \omega \leq \omega_{RC} \tag{3.1}$$

The proposed robust controller and FDRRC model is shown in Figure 3.3. In the present study, a second order FDRRC model has been considered with two FDRC loops.

The proposed second order repetitive controller has the following transfer function:

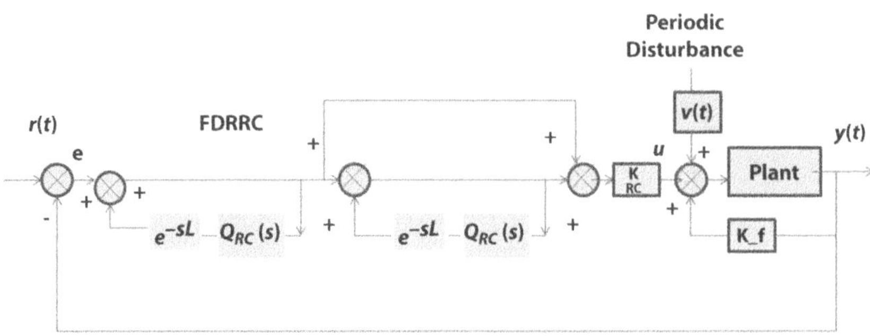

Figure 3.3 Proposed FDRRC model.

$$G_{FDRRC}(S) = \frac{yc(s)}{e(s)} = \frac{1-(1-Q_{RC}(s)eL)^2}{(1-Q_{RC}(s)e^{-sL})^2}e^{sL} \tag{3.2}$$

3.3 Problem Formulation

Overall, a closed loop system block diagram considering input disturbance is given in Figure 3.4. The objective is to choose the proposed controller parameters in a robust way so that the fundamental frequency and harmonic frequency components of the periodic disturbance signal are attenuated asymptotically. The fundamental frequency component of the disturbance and all its harmonics are rejected asymptotically.

The FDRRC gain parameter K_{RC} regulates the repetitive control action to be applied. It also affects the stability. Choosing a higher gain value may result in fast system response for disturbance rejection but it can make the overall system unstable. There must be a trade off between convergence of tracking error and overall system stability. Therefore, the right gain value must be chosen to satisfy robust stabilization criteria which has been discussed in the later sections of this chapter.

In this present study, it is assumed that the periodic disturbance is considered as an input disturbance $v(s)$ to the plant.

The robust characteristics to suppress or reject disturbance with respect to error can be defined with the sensitivity function (S_{di}) of the FDRRC system. The expression for the function is given by:

$$S_{di} = \frac{e(s)}{v(s)} = \frac{-1}{1+G_P(s)\times G_{FDRRC}(s)} \tag{3.3}$$

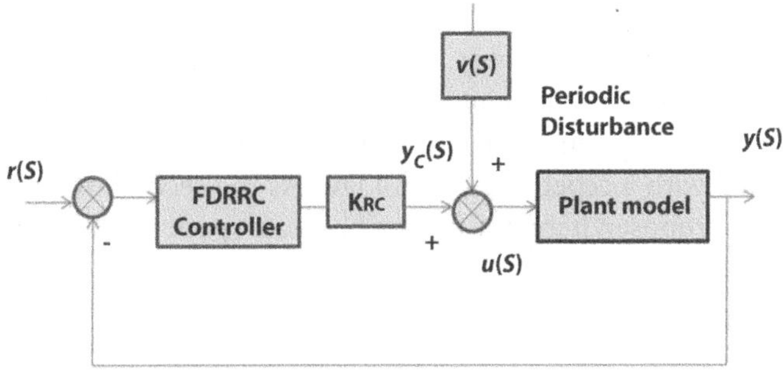

Figure 3.4 Closed loop system block diagram.

Two different cases of periodic disturbance rejection can be there: fixed frequency and varying frequency disturbance signal. Here, in the present study, the fixed frequency known and unknown periodic disturbance rejection is addressed. The disturbance signal can be represented as a summation of sinusoids, considering first and second harmonics.

$$v(t)=\sum a_1 \sin(wt)+a_2 \sin(2wt) \tag{3.4}$$

The disturbance signal along with period variation can be represented as:

$$v_{new}(t)=\sum a_1 Sin(\frac{2\pi t}{L+\Delta L})+a_2 \sin(\frac{4\pi t}{L+\Delta L}) \tag{3.5}$$

where ΔL is amount of variation.

3.4 LMI Formulation for Robust Stabilization Criteria

In FDRC state space formulation cut off frequency of the low pass filter acts as an elemental part as the controller states are found from it.

Equivalent state space model of the FDRC loop is given by:

$$\dot{X}_{RC'}(t)=-\omega_{RC}X_{RC'}(t)+\omega_{RC}X_{RC'}(t-L)+\omega_{RC}e(t-L) \tag{3.6}$$

$$Y_{RC'}=X_{RC'}(t)+e(t) \tag{3.7}$$

Vector $X_{RC'}$, $Y_{RC'}(t)$, $e(t)$ are the state variable, controller output and error vector, respectively.

This work presents a robust method to actively reject periodic disturbances with the implementation of a multiloop finite dimensional repetitive-control system where the input periodic disturbance will be compensated with a repetitive control law. As the second order repetitive controller is proposed here, the equivalent state space model is given by:

$$\dot{X}_{RC'}=-\omega_{RC}x_{RC'}(t)+\omega_{RC}x_{RC'}(t-L)+\omega_{RC}x_{RC'}(t-2L)+e(t-L)+e(t-2L) \tag{3.8}$$

In state space formulation, A_C, B_C, C_C, and D_C are the system matrices of FDRRC.

It is given by the following state equation.

$$\begin{aligned} \dot{X}_{RC'} &= A_C x_{RC'}(t) + B_C C_{RC}(t) \\ Y_{RC'}(t) &= C_C x_{RC'}(t) + D_C Y(t) \end{aligned} \tag{3.9}$$

The augmented plant model along with controller state and disturbance state is formulated. The augmneted system is given by:

$$\dot{X}_a(t) = AX_a(t) + A_{td}X_a(t-L) + Bu(t) \tag{3.10}$$

$$X_a(t) = [X_P(t)^T \quad X_{RC'}(t)^T]^T \tag{3.11}$$

$$u(t) = K_{RC}r(t) + KX_a(t) \tag{3.12}$$

$$K_{new} = [(K_f - K_{RC}C) \quad K_{RC}] \tag{3.13}$$

To obtain linear-matrix-inequality-(LMI)-based criterion for stability, a Lyapunov Krasovskii function is considered. One tuning parameter is introduced in the LMI, which can be manipulated and the robust stability can be adjusted. With this, the learning efficiency of the controller will be improved.

Filter cut off frequency in FDRC loops, ω_{RC} is a known parameter. The robust stability condition is obtained with the Lyapunov Krasovskii functional V'(t). According to argument of stability principle the reference signal is made zero to obtain the gain matrix K, given the value of ωRC.

$$V'(t) = X_a^T(t)PX_a(t) + \int_{t-L}^{t} X_a^T(\eta)QX_a(\eta)d\eta \tag{3.14}$$

Lyapunov Krasovskii theory for stability says that $\dot{V}'(t) < 0$, given $P = P^T > 0$ and $Q = Q^T > 0$

The Lyapunov Krasovskii function derivative is given as:

$$\dot{V}'(t) = \dot{X}_a{}^T(t)PX_a(t) + X_a{}^T(t)P\dot{X}_a(t) + X_a{}^T(t)QX_a(t) - X_a{}^T(t-L)QX_a(t-L) \tag{3.15}$$

$$\begin{aligned}\dot{V}'(t) = {} & X_a{}^T(t)A^T PX_a(t) + X_a{}^T(t)PAX_a(t) + X_a{}^T(t)PA_{td}X_a(t-L) + X_a{}^T(t)PBu(t) \\ & + X_a{}^T(t)QX_a(t) - X_a{}^T(t-L)QX_a(t-L)\end{aligned} \tag{3.16}$$

$$\begin{aligned}\dot{V}'(t) = {} & X_a{}^T(t)A^T PX_a(t) + X_a{}^T(t)PAX_a(t) + X_a{}^T(t)PA_{td}X_a(t-L) + X_a{}^T(t)PBK_{RC}r(t) \\ & + X_a{}^T(t)PBKX_a(t) + X_a{}^T(t)QX_a(t) - X_a{}^T(t-L)QX_a(t-L)\end{aligned} \tag{3.17}$$

$$\begin{aligned}\dot{V}'(t) = {} & X_a{}^T(t)P(A+BK)X_a(t) + X_a{}^T(t)PA_{td}X_a(t-L) + X_a{}^T(t)PBK_{RC}r(t) + X_a{}^T(t)QX_a(t) \\ & - X_a{}^T(t-L)QX_a(t-L)\end{aligned} \tag{3.18}$$

Some algebraic manipulation of the above expression results in:

$$\begin{bmatrix} X_a(t) \\ X_a(t-L) \end{bmatrix}^T \Theta \begin{bmatrix} X_a(t) \\ X_a(t-L) \end{bmatrix} < 0 \tag{3.19}$$

where

$$\Theta = \begin{bmatrix} (A+BK)^T P + P(A+BK) + Q & PA_{td} \\ A_{td}{}^T P & -Q \end{bmatrix} \tag{3.20}$$

Here, $\Theta < 0$ meets the condition given by Equation 3.20.

Both sides of above matrix inequality is multiplied by P^{-1} and by replacing

$X_a = P^{-1}$, $Y = KX_f$ and $S = P^{-1}QP^{-1}$, the following matrix inequality is derived. (3.21)

$$\Theta = \begin{bmatrix} (AX_a + BY + X_aA^T + Y^TB^T + S & A_{td}X_a \\ X_aA_{td}^T & -S \end{bmatrix} < 0 \quad (3.22)$$

3.5 Plant Model

In literature, many rotary applications have been mentioned. To validate the performance of the proposed controller, two different example cases of DC servo motor and rotary peristaltic pump are taken. Modelling of this system is shown here briefly. For both the cases, the constant rotational speed of the motor and the rotational frequency is substantially constant. Sometimes, a kind of unevenness is noticed in rotational speed and a significant amount of fluctuation is found in rotational speed. One of the major reasons is the presence of a disturbance in the system. Therefore, robustness in the vicinity of the rotation speed control is required with the periodic disturbance compensation.

a) *Test Case 1: DC Servo Motor*

Automation has become a very important part of different engineering applications. The vehicle control system, motion control system, and also robotic control systems involve an extensive use of the same. A DC servo motor is one of the main units of automatic control system modules and several drive application modules that are used in the industry nowadays. It has better controllability characteristics and flexibility in different types of automatic control applications as mentioned earlier. Therefore, the precise position control of the DC servo motor is of major control interest, hence the disturbance compensation [23]. In Figure 3.5, the equivalent electrical parameter model block diagram of the DC servomotor is provided. For precise angular speed control, rejection of disturbances is the control objective here. All the parameters are listed in Table 3.1.

The permanent magnet DC motor used to actuate the servo system is a permanent magnet type. The dynamic model can be represented with the following differential expressions. The transient response of the motor is represented with the given equations below.

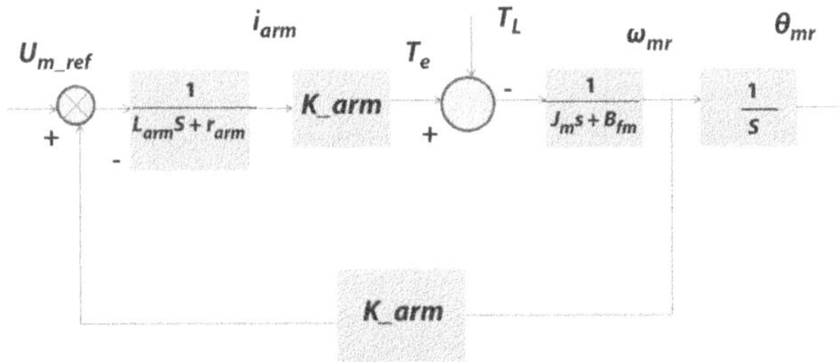

Figure 3.5 Block diagram of DC servo motor.

Table 3.1 DC servomotor parameters.

Symbol	Parameters	Symbol	Parameters
U_{m_ref}	Motor armature voltage	θ_{mr}	Angular position of motor rotor
L_{arm}	Motor inductance	ω_{mr}	Measured rotational angular speed
r_{arm}	Armature resistance	J_m	Moment of inertia of rotor
$K_{\_arm}$	Electromotive force constant	B_{fm}	Damping ratio of motor
i_{arm}	Armature current	T_L	Torque of motor load
d_m	Amplification factor of duty ratio to control the armature voltage		

$$\frac{du_{m\_ref}}{dt} = -\frac{1}{T_i}u_{m\_ref} + \frac{K_i}{T_i}d_m \tag{3.22}$$

$$\frac{di_{arm}}{dt} = \frac{1}{L_{arm}}u_{m\_ref} - \frac{R_{arm}}{L_{arm}}i_{arm} - \frac{K_{arm}}{L_{arm}}\omega_{mr} \tag{3.23}$$

$$\frac{d\omega_{mr}}{dt} = \frac{K_{arm}}{J_m}i_{arm} - \frac{B_{fm}}{L_{arm}}\omega_{mr} - \frac{1}{J_m}T_L \tag{3.24}$$

$$\frac{d\theta_{mr}}{dt} = \omega_{mr} \tag{3.25}$$

The equivalent state space model can be obtained for the DC servo motor from the above differential equation. It is given by:

$$\begin{bmatrix} \dot{u}_{m\_ref} \\ \dot{i}_{arm} \\ \dot{\omega}_{mr} \\ \dot{\theta}_{mt} \end{bmatrix} = \begin{bmatrix} -\dfrac{1}{T_i} & 0 & 0 & 0 \\ \dfrac{1}{L_{arm}} & -\dfrac{R_{arm}}{L_{arm}} & -\dfrac{K_{arm}}{L_{arm}} & 0 \\ 0 & \dfrac{K_{arm}}{J_{fm}} & -\dfrac{B_{fm}}{J_m} & 0 \\ 0 & 0 & 1 & 0 \end{bmatrix} \begin{bmatrix} u_{m\_ref} \\ i_{arm} \\ \omega_{mr} \\ \theta_{mt} \end{bmatrix} + \begin{bmatrix} \dfrac{K_i}{T_i} \\ 0 \\ 0 \\ 0 \end{bmatrix} d_m + \begin{bmatrix} 0 \\ 0 \\ \dfrac{1}{J_m} \\ 0 \end{bmatrix} T_L \tag{3.26}$$

b) *Test Case 2: Rotary Peristaltic Pump*

Multipurpose application of rotary peristaltic pump is found since 1930's [24]. Rotary roller pumps are available in variable configurations depending upon the number of rollers. Accordingly, the pump is to be modelled and equivalent electrical model of the pump can be derived.

A flow centric equivalent electrical model is referred to here [25]. The model has undergone certain assumptions that at least one roller will always be attached with the tube and there will be full occlusion during the span of attachment and the process will be frictionless. Figure 3.6 shows the equivalent electrical model of the pump. Current voltage relationship, which is analogous to the pump pressure and flow rate is given by the governing equations at inlet and outlet node 1 and 2 by applying KVL. All the equivalent electrical parameters are given in Table 3.2.

$$P - P\_in(t) = R\_in \times M(l)\_in(t) + L\_in \frac{dM(l)\_in(t)}{dt} \tag{3.27}$$

$$P\_out(t) - P = R\_oyt \times M(l)\_out(t) + L\_out \frac{dM(l)\_out(t)}{dt} \tag{3.28}$$

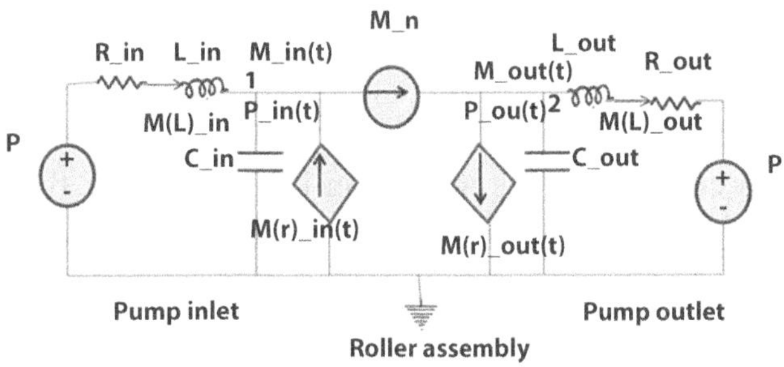

Figure 3.6 Equivalent electrical model of peristaltic pump of flow centric approach.

Table 3.2 Equivalent electrical model parameters of rotary peristaltic pump.

Pump parameters	Equivalent electrical component
Fluid Friction	Resistance R
Compliance	Capacitance C
Fluid Inertia	Inductance L
Pressure	Voltage Quantity P
Volumetric Flow Rate	Current Quantity M

State space matrices for the both the models are provided in the next section.

3.6 Simulation Study

1V step reference signal is taken here for DC servomotor speed and peristaltic pump flow rate control. The system matrices for DC servo motor and rotary peristaltic pump are given below.

State matrices for DC servo motor:

$$A_{dcm} = \begin{bmatrix} 10000 & 0 & 0 & 0 \\ 250 & -675 & -25 & 0 \\ 0 & -1000 & -0.8 & 0 \\ 0 & 0 & 1 & 0 \end{bmatrix}, B_{dcm} = \begin{bmatrix} 30000 \\ 0 \\ 0 \\ 0 \end{bmatrix}, C_{dcm} = \begin{bmatrix} 0 & 0 & 0 & 1 \end{bmatrix}$$

State matrices for rotary pump:

$$A_{pp} = \begin{bmatrix} -2.6040 & -5.2710 \\ 1 & 0 \end{bmatrix}, B_{pp} = \begin{bmatrix} 1 \\ 0 \end{bmatrix}, C_{pp} = \begin{bmatrix} 0 & 4.0180 \end{bmatrix}$$

FDRRC, delay parameter(L) and filter cut off frequency (ω_{RC}) are chosen as 2 seconds and 10 rad/sec respectively to provide the sufficient controller bandwidth. Gain parameters of the proposed repetitive control law for a robustly stable feedback system is designed using LMI condition as derived in earlier section.

Following two disturbance sinusoids given by $v_1(t)$ and $v_2(t)$, shown in Figure 3.7 (a) and (b) are considered for simulation case study $v_1(t)$=0.6 (0.2πt).

$$v_2(t) = 0.5 \sin(0.5\Pi t) + 0.5 \sin(\Pi t)$$

The gain matrix K_1 for the DC servo motor is computed from Equation 3.21.

$$K_1 = \begin{bmatrix} -2.89\times 10^{10} & -2.89\times 10^{12} & 0 & -1.0002 & 2.89\times 10^{10} \end{bmatrix}$$

The gain matrix K_2 for the rotary pump is computed from Equation 3.21.

$$K_2 = \begin{bmatrix} -2.38\times 10^{12} & -1.0056 & 2.38\times 10^{12} \end{bmatrix}$$

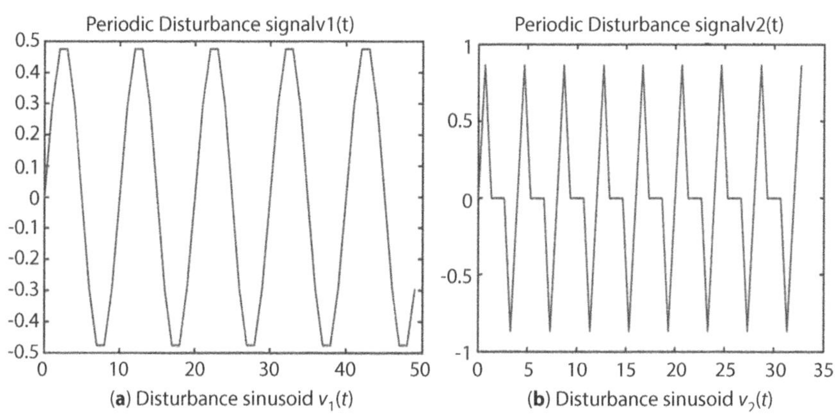

Figure 3.7 Disturbance sinusoids.

By looking into the tracking error convergence plot, in Figures 3.8 and 3.9 for two different rotary systems which involve periodic disturbance, it can be said that the proposed FDRRC controller is performing significantly better in disturbance attenuation compared to conventional repetitive controller.

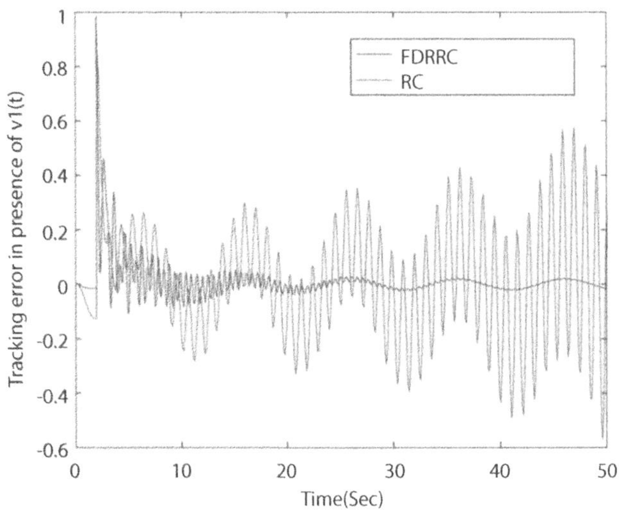

Figure 3.8 Tracking error convergence with FDRRC and RC for peristaltic pump.

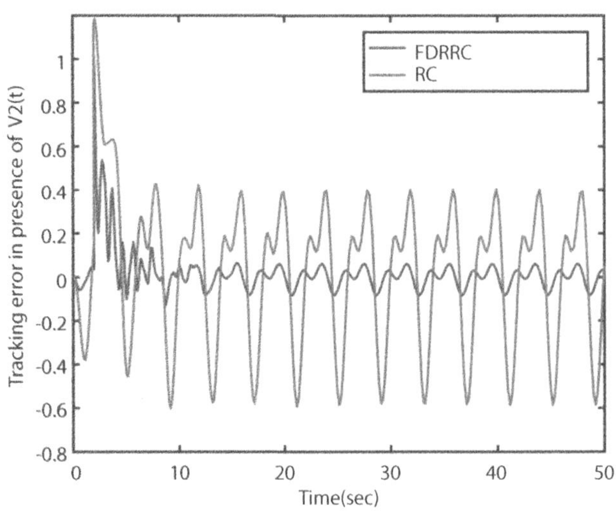

Figure 3.9 Tracking error convergence with FDRRC and RC for DC servo motor.

3.7 Processor in Loop (PIL) Simulation

Traditional simulation procedures cannot always perform to the exact control dynamics. Processor-in-loop (PIL) simulation is a method to bridge the gap between simulation study and real time system dynamics. Once the control algorithm is developed, then it can be embedded to a hardware platform [26–28]. The software execution in real time cannot be examined in the PIL, but it is permitted to verify whether the proposed controller firmware can run in a standalone processor or not.

In this present study, the proposed controller subsystem has been designed in a simulation environment. Later, it was tested using a real time processor-in-loop simulation. A Raspberry Pi 3B model microcomputer has been used for the same. This real time simulation provides access to the hardware controller.

The RTPIL simulation procedure is a good alternative to the simulated control system having a real time hardware controller. It makes the proposed design more reliable. In the future, this work can be further extended to hardware-in-loop simulations to validate the design with real time plant applications. Figure 3.10 shows the block diagram of the overall method. Figure 3.11 shows the real time set up. All the simulation has been carried out using MATLAB SIMULINK. Later, the SIMULINK model was embedded to the Raspberry Pi module. The steps of execution are listed below.

- Step 1: Connectivity configuration setup is to be done on the PC where the simulation software (MATLAB in this present work) is running
- Step 2: To set up the communication, the PC needs to be connected with Raspberry Pi using serial communication protocol
- Step 3: The control algorithm (SIMULINK model) is to be downloaded to the target processor (Raspberry Pi) via a serial connection link
- Step 4: Target processor sends a response to the PC once the firmware is loaded into it

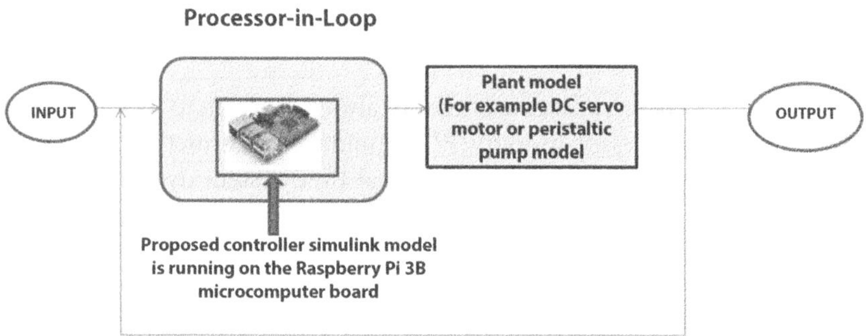

Figure 3.10 Block diagram of processor in loop simulation of proposed controller.

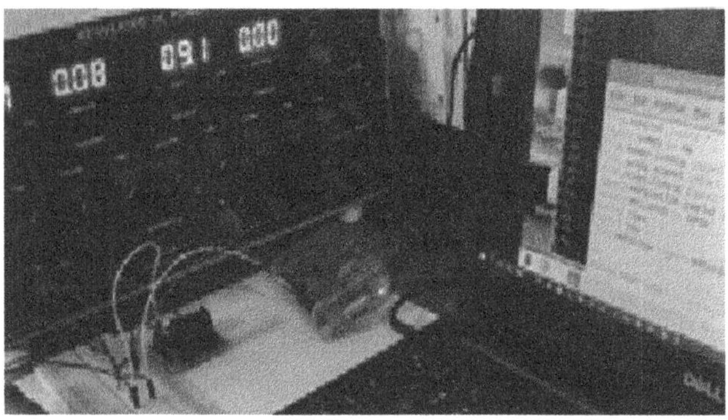

Figure 3.11 Real time set of processor in loop simulation of proposed controller.

3.8 Conclusion

This chapter provides some important insights about periodic disturbance rejection in rotary bodies. Different control methodologies have been investigated. A finite dimensional robust repetitive controller (FDRRC) compensates for periodic disturbances in the rotary system. The fixed frequency disturbance signal is considered here. The proposed controller with two tuning parameters provides a simpler solution to the input disturbance factor to the plant model. As the proposed method offers a higher order controller design, the design procedure ensures the robust system stability. LMI based convergence criteria is obtained. Two different plant models of the DC servo motor and rotary three roller pump is taken here as an example case as they are widely used rotary systems in many industrial applications. Two different disturbance signals are taken during simulation and it is shown that the proposed controller can compensate periodic disturbances for measurably accurate tracking performance compared to a conventional controller model. Comparative simulation studies are shown. A processor in loop (PIL) simulation is also carried out using a Raspberry Pi 3B board to extend the simulation study on a more realistic level. Controller design can be modified further considering a quadratic cost function and formulated as an optimization problem for controller parameter design for varying frequency input disturbances.

References

1. Lee, H.S. and Ryu, S. (2021). Design of a robust controller for a rotary motion control system: disturbance compensation approach. Microsystem Technologies, 2293–2302
2. Sacks, A., Bodson, M., & Khosla, P. (1996). Experimental Results of Adaptive Periodic Disturbance Cancellation in a High Performance Magnetic Disk Drive. ASME Journal of Dynamic Systems, Measurement, and Control, 118, 416-424.
3. Bittanti, S. & Moiraghi, L. (1994). Active Control of Vibrations in Helicopters via Pole Assignment Techniques. IEEE Transaction on Control Systems Technology, 24, 343-350.
4. Huang, Y. & Messner, W. (1998). A novel disturbance observer design for magnetic hard drive servo system with a rotary actuator. IEEE Transactions on Magnetics, 34, 4, 1892-1894.
5. Kosaka, M. and Uda, H. (2010). Generalised Disturbance Observer for Systems with Harmonic Disturbances and Input Saturation Using the Internal Model Principle. Journal of Low Frequency Noise, Vibration and Active Control, 29(1).55-64.
6. Seo, H.T., Kim, K.S. and Kim, S. (2015). Add-on controller design based on disturbance observer in H-infinity framework. IEEE Digital Explore. 15th International Conference on Control, Automation and Systems, 282-285.
7. Dey, N., Mondal, U. and Sengupta, A. (2019). Modified Repetitive Controller for Disturbance Rejection of Peristaltic Pump. IEEE Digital Explore. International Conference on Opto-Electronics and Applied Optics, 1-5.
8. Ramos, G. A., Castello, R.C. and Olm, J.M. (2013). Digital repetitive under varying frequency conditions. Springer, 1-159.
9. Kurniawan, E., Cao, Z. and Man, Z. (2014). Design of robust repetitive control with time-varying sampling periods. IEEE Transactions on Industrial Electronics, 61, 6, 2834–2841.
10. Pipeleers, G., Demeulenaere, B., Schutter, J.D. and Swevers, J. (2008). Robust high-order repetitive control. IEEE American Control Conference, 1080–1085.

11. Wu, Y. & Yue, D. (2018). Desired compensation adaptive repetitive control of electrical-optical gyro-stabilized platform with high precision disturbance compensation. Asian Journal of Control, 20, 2, 867–880.
12. Huo, X., Wang, M., Liu, K. & X. Tong, X. (2020). Attenuation of Position-Dependent Periodic Disturbance for Rotary Machines by Improved Spatial Repetitive Control With Frequency Alignment. IEEE/ASME Transactions on Mechatronics, 25,1, 339-348.
13. Steinbuch, M. (2000). Repetitive Control for Systems with Uncertain Period-Time. Automatica,38, 2103-2109.
14. Jamil, M., Rashid, U., Arshad, R., Khan, S., Gilani, S.O. & Ayaz,Y. (2015). Robust Repetitive Current Control of Two-Level Utility-Connected Converter using LCL Filter. Arabian Journal for Science and Engineering, 40, 2653–2670 .
15. Jamil, M., Waris, A., Gilani, S.O., Khawaja, B. A., M. N. Khan, M. N. & Raza, A. (2020). Design of Robust Higher-Order Repetitive Controller Using Phase Lead Compensator. IEEE Access, 8, 30603-30614.
16. Guo, H., Otomo, K. & Ishihara, T.(2020). A Design Method of High Order Repetitive Controllers. IFAC-Papers On Line, 53,2, 4440-4445, 2405-8963.
17. Flores, J.V., Salton, A.T. & Castro, R.S. (2016). Robust State Feedback Formulation for High Order Repetitive Controllers. Asian Journal of Control, 18, 3, 1042–1051.
18. Mondal, U., Sengupta, A. and Pathak, R.R. (2016). Servomechanism for periodic reference input: Discrete wavelet transform-based repetitive controller. Transactions of the Institute of Measurement and Control, 38(1)14-22.
19. Zhou, L., She, J., Wu, M. and He, Y. (2010). LMI-based design method of robust modified repetitive-control systems. IEEE ICCA, 440-445.
20. Sangeetha, G.R. & Jacob,J. (2008). Repetitive controller for periodic disturbance rejection in motor-gear transmission system. Annual IEEE India Conference, 559-564.
21. Steinbucha, M., Weiland, S. and Singh, T. (2007). Design of noise and period-time robust high-order repetitive control with application to optical storage. Automatica, 43, 2086-2095.
22. Singh,T. & Vadali, S.R. (1993). Robust Time-Delay Control. Journal of Dynamic Systems, Measurement and Control, Transactions of the ASME, 115, 303-306.
23. Wang, Y., Hu, X., Feng, Y. and Li, S. (2020). Design and Experimental Study of DC Servomotor Controller. 3rd International Conference on Advanced Electronic Materials, Computers and Software Engineering, 684-689.
24. Fei, D., Deng, C., Zhang, Q. and HUNAG, Z. (2015). Repetitive Control to the Flow Rate of Peristaltic Blood Pumps. International Advanced Research Journal in Science, Engineering and Technology, 2, 8, 1-5.
25. Maré, L.D., Velut, Hagander, S.P. and Akesson,M. (2001). Feedback control of flow rate from a peristaltic pump using balance measurements. European Control Conference, 2164-2168.

26. Bacic, M. (2005). On hardware-in-the-loop simulation. Proceedings of the 44th IEEE Conference on Decision and Control, 3194-3198.
27. Jiang, Z., Leonard, R., Dougal, R., Figueroa, H. and Monti, A. (2004). Processor-in-the-loop simulation, real-time hardware-in-the-loop testing, and hardware validation of a digitally-controlled, fuel-cell powered battery-charging station. IEEE 35th Annual Power Electronics Specialists Conference, 3, 2251-2257.
28. Liu, R., Monti, A., Francis, G., Burgos, R., Wang, F. and Boroyevich, D. (2007). Implementing a Processor-in-the-loop with a Universal Controller in the Virtual Test Bed. IEEE Power Electronics Specialists Conference, 945-950.

4

Proctored Secure Face Lock System

Kalpana Murugan*, Sathya Pradeep C., Krishnapriya J. and Gayatri P.

Department of Electronics and Communication Engineering, Kalasalingam Academy of Research and Education, Virudhunagar, Tamil Nadu, India

Abstract

Securing our home with high-tech equipment is a wish and need for all of us in this society. Smartphones are of great advantage in this aspect. Using this electronic device allows us to remotely access and control equipment and appliances kept in the home or office from anywhere in the world, as an added advantage. To serve this purpose, many technologies are invented. One of the effective technologies is wireless technology among the various other technologies invented because accessing the gadgets connected to this network is always easy. The proposed work creates a facial recognition-based shutter-based mechanism using raspberry pi. This device sends an "intruder message" or "owner message" to the owner's mail ID if the house is accessed by an intruder. By sending messages through the mail, privacy is maintained and enables only the owner to act. The device takes photos/snaps of the person in front of the camera positioned in the shutter locks and through a python code they are processed as mentioned earlier. The shutter mechanism is controlled by the servo motor in the prototype made. The motor rotates 90 degrees for about a certain delay and then closes automatically by again rotating back 90 degrees in the opposite direction. Meanwhile, the mail will be sent with the photo of the intruder. The device relates to housing Wi-Fi for sending mail. The facial recognition-based shutter lock and mechanism using raspberry pi can be used in various places such as banks, IT companies, government offices, military, etc.

Keywords: Raspberry PI, PI camera, PIR sensor, house automation, facial recognition, shutter mechanism

**Corresponding author*: drmkalpanaece@gmail.com

Vipin Chandra Pal, Suman Lata Tripathi, and Souvik Ganguli (eds.) Industrial Control Systems, (59–76) © 2024 Scrivener Publishing LLC

4.1 Introduction

In today's technological era, various electronic equipment plays a very important role in our day-to-day routine. Each and every advancement in electronic technology affects our routine through the tiny device which settles inside our palm, i.e., our smartphones. They contribute to and affect our lives and, moreover, have become a part of life. They help us in computations and adjustments for accomplishing such technologies. Protecting our homes is one of the important issues that is faced by people when they lock up their houses and leave valuable things like gadgets, money, or even kids and grandparents that must be kept safe inside the home. Appointing a human to guard them can be expensive and cannot provide 24/7 security. So, digitally inventing a device to secure our homes will be a solution to this problem. Home automation is a technique that turns the appliances in the home into smart automation only for the sake of monitoring them remotely. For monitoring remotely, home automation is a system in which the central controller is connected to various components or gadgets that are intended to be controlled either wired, in case of older versions, or wirelessly. The older variants were a bit tedious because they required a good pre-plan before implementation and adding any new feature led to rearranging the whole previous structure. So obviously, the best option emerged to be the wireless method and compromised is the cost too. The first thing that must be secured in the home is the door, so using an electronic lock working with IoT provides a secure system that has an interactive interface to authorize the user to get permission. In this aspect, a concept called NFC (Near Field Communication) was widely used for lock mechanisms using facial points.

A wireless surveying and monitoring door/shutter lock mechanism is developed like a prototype model. Raspberry Pi is a handy and comparatively small computer that can function almost like a computer. In the aspect of surveillance, many inventions have taken place in the past and are still taking place. Some of the inventions are the surveillance camera, CCTV, etc. In these types of surveillance systems, a person is always required to monitor the things going on and cannot be done remotely. But, in our project, wherever the user may be, either on station or off station, he/she can keep an eye on what is happening in the place where the device is installed. Moreover, in this method of privacy the person on both sides is maintained as it is viewed and controlled by only one user. All these advantages can be done by using an easy raspberry circuit. Some devices can work in their own operating systems, likewise, raspberries have their own operating system.

4.1.1 The Need for Technology

The security of lockers or home doors is always a concern. As the technology grows day by day, many security systems are available for switching ON and OFF the appliances and automation of appliances. The most familiar modes of locking mechanism are a fingerprint locking mechanism, iris mode of locking mechanism, and various other modes. Every mode of authentication has its pros and cons with the locking system. Many locking mechanisms might be available but the best fit in terms of security should be something which enables only the owner to access it and when there is a security breach, the owner should get the notification on it through the best mode, either a mail or mobile application. This application not only safeguards us, but also provides authentication to open or shut the door according to the owner's command. So, no other person is required to safeguard the house if the owner is out of the station. In another case, malpractice is also not possible even if the owner is at the house at night.

4.2 Background

The system which has been designed using Raspberry Pi 3B+ model as shown in Figures 4.1 and 4.2 which has two main parts: Controlling and Monitoring. The designed system uses raspberry pi as a controller and an ultrasonic sensor for monitoring (Manoj *et al.*, 2020). A CCTV camera has been used to monitor the person or object that falls in front of the camera and a GSM Module is used to send messages to an authorized person about the unauthorized person trying to unlock the door (Agarwal and Kumar, 2019). The microcontroller is used to connect the appliances with others and the PIR sensor is used for monitoring people. The message will be sent using a cryptographic algorithm so that the message will be authenticated (Ahtsham *et al.*, 2019). The Viola-Jones Face Detection method is used to extract similar features from the database provided. The above-mentioned GSM module is also used to send messages. The servo motor mechanism is utilized to unlock and lock the door (Nag *et al.*, 2018). The algorithm used to differentiate features between images is the Viola-Jones Face detection method. The USB camera module is used to capture images of persons (Sandesh Kulkarni *et al.*, 2017). Face recognition uses the support vector machine (SVM) which is utilized to train the database that has been provided in the cloud and local binary patterns (LBP) are done using gradient-vector and histogram. The ESP-32 camera is used to detect the images (Jiakailin Wang *et al.*, 2020). The study has been done in the smart home security systems and most of the proposed work uses GSM-GPRS protocols, GSM modem, and wireless network monitoring using Wi-Fi. The work is done especially for home automation (Chitnis *et al.*, 2016).

In this paper, a system is designed so that it can make use of the sensors in detecting movement, intensive heat, noise, and fumes which are controlled by a PIC microcontroller 16F76. The detected information from the sensor is collected, then the system decides whether to send the short message notification to the user's registered mobile number with the help of the GSM module (Lova Raju *et al.*, 2019). If any kind of disturbance signal is recorded from the sensor, then the PIC system decides to send an SMS to the user's registered mobile number and the nearby police station for security measures. Comparably, if it detects any smoke and fire-like threats in the surrounding area, then it sends an SMS to the user and the nearby fire service station. SMS can only be sent to the user's number only if the GSM module is connected to the system and the SMS service is well used in this system to make the monitoring more effective (Kim and Ban, 2015). In this, the proposed system helps in tracking and detecting the invader around the system's surroundings and is of the light-sensitive model. A face recognition system is also used so that the image of the invader can then be sent to the user's e-mail for monitoring purposes and additionally helps in identifying and detecting the invader. To distinguish between various objects, the system was inbuilt with an algorithm like tracking or movement. Unlocking passwords is not used in this system.

Based on the analysis made on security systems, the preexisting technologies are different from face detection, in which most of them were touch-based (biometric), swipe-based (RFID cards), or a GSM-based Time Password. Each of these technologies is individually sound in certain areas but lags in providing total security. Plenty of researchers have researched home automation and security systems, but only countable numbers of them did that using face recognition. Face recognition-based security systems offer good potential security for household gadgets if used (Karimi, 2020). The next paper explains how LBP works with the face detection module and how it helps train the raspberry pi but lacks support during a power loss situation.

The training of a microcontroller fixed to the door's unlock-lock feature for facial recognition is done using HAAR transformation, which is one of the oldest facial recognition methods and ineffective for the current trend. It can even detect a 2-D photograph of a face, but this would provide a very wort result with the security concept. In this paper, Wi-Fi limits its area of coverage and the device user has no option to access or monitor it remotely (Begum, 2019). Another major disadvantage of the above systems for the current trend and era is that everyone uses a smartphone and users like the concept of monitoring their devices by using them. But, these systems either do not inform the user regarding the security breach issue or use a separate module just to do that work. By adding these kinds of extra burdens, the cost increases, power consumption also increases, and the user gets no chance of acting in case of a security breach.

Facial recognition algorithms started emerging in this world in the early' 90s. Since then, it has been of great use in various places. It is used to identify a person's unique mark or identification and then compare it with the database for similarity. Support Vector Machine (SVM), Viola-Jones (VJ) Algorithm, and Kanade Lucas Tomasi Algorithm are some the algorithms which are widely preferred and used by the researchers in today's world (Jana Alghamdi *et al.*, 2021). Facial recognition has four basic steps which are face detection, feature extraction, training with the HAAR, and then face recognition. Face detection also has four steps, namely image acquisition, histogram normalization, noise removal, and face detection (Yadav *et al.*, 2018).

In the case of deep learning, the CNN are an effective method to scan the face and perform facial recognition. In order to train convolutional neural networks (CNN), plenty of data is required which might be a disadvantage. To overcome this advantage, one of the methods is to use Face Net, which is a google LLC facial recognition and clustering function. This method provides an accuracy of 99.63 percent (Asra, 2019). In (Wang *et al.*, 2014) and the paper gives an idea about using the facial recognition method to create a device that can get the attendance of people in real-time. This method helped in identifying the tool to do the process and gives a briefing on how to create such devices. This helped identify if our idea would work.

Speaking about the door lock mechanism, there are various methods of electronic locks available which can be made by working with a simple microcontroller. Some of the systems are a Password Lock System, RFID Lock System, Biometric Lock System, and Cryptography based Lock System. These types of lock mechanism ideas gave us the thought of using a new method called facial recognition (Ghosh, 2021 and Divya and Meera Mathew, 2017).

Usage of an electronic lock system asks for some data like passwords or unique data to unlock. This is a better mechanism than the normal key lock variant (Happy N. Monday *et al.*, 2019). The (Amuda and Tennyson, 2017) and (Anusha *et al.*, 2018) paper helped us to understand how data acquisition works for the case of a lock. The paper helped us understand how data is to be stored first and then, using the data, what occurred at that moment to compare with the pre-existing thing. A similar ideology as stated in the paper (Prayag Bhatia *et al.*, 2020) is utilized for creating a door lock mechanism and to use the ideology of training the facial recognition algorithms using the LBP histogram method. The feature vectors automatically created by the Histogram of Gradients descriptor are used to train Support Vector Machines (SVM) and the results are compared with the test inputs that are given to the device (Dong *et al.*, 2021). Some of the similar methods use the Viola-Jones method to detect faces and the Eigenfaces method to recognize faces and run the comparison code (Fahim Faisal and Syed Akhter Hossain, 2020). Some methods use social media networks to send the message of the intruder instantly, but the problem is that it cannot give control back to the device, the message cannot be encrypted for security purposes and privacy, and issues may occur (Hema and Juli Yadav, 2020).

4.3 Proctored Secure Face Lock System

The designed smart home security system using facial recognition is carried out by connecting with raspberry pi as shown in Figure 4.1. Due to the fast-processing nature of raspberry pi, it has been used. Every system follows these three steps: getting input from the pi camera and database, processing the designed system to differentiate features between the captured image and provided database, and confirming the door lock and unlock mechanism. The input step focuses on getting input from the pi camera, which shows the person or object that falls in the allocated area of the home. The camera matches the captured image with the database using the Local Binary Pattern (LBP) algorithm. If the intruder or any person's image corresponds with the input database, then the door will be unlocked by a shutter mechanism. If the person does not match the database, then the image from the pi camera will be sent to the owner by mail/telegram application. If the owner wants to provide permission to the person standing in front of the camera, then the owner can click the appropriate option to permit. The door will be unlocked after that and stores the new person image in the database.

Communication is done by using Raspberry PI and email, where raspberry pi is a system that can be connected to various modules to carry out the desired operation and email is the communication sent to the owner from the pi camera and trained algorithm. The input module is the PI camera in this work, which captures the picture of the intruder that appears in front of the camera. The control module is the raspberry pi module which has been connected to the PI camera and PIR sensor which detect the motion and faces of the persons. It is the major controllable module which

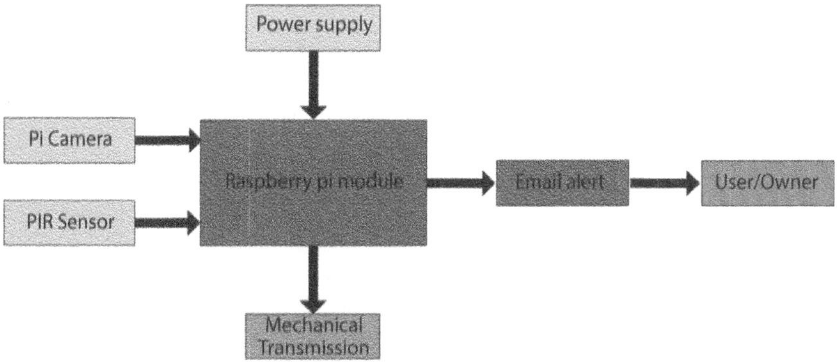

Figure 4.1 Block diagram of lock system.

either allows the person or does not permit the person which is significant for the safety purpose of the home.

4.3.1 Methodology

At first, initialization is carried out and the power supply is fed to the raspberry pi module. It is connected with the battery system to power the internal components. The device's internal components are subjected to initialization in order for all of the electronic components to function properly. The PIR Sensor remains ON to detect motion once startup is complete, while all other components are turned off. After initialization, the PIR detects some motion and triggers the camera ON through raspberry pi. The camera senses whether the object standing in front of it has a human face or not (done by the pre-training given to the camera module through the raspberry pi module). Then, Detection and Authentication are carried out and once the pi camera detects the face, it takes a snap of it. Then the snap is compared with the database. An authentication step is done with the database to confirm whether the face of the intruded is the owner's face or a stranger. This authentication is carried by the raspberry pi module. While sending the status, if the right match is found after running the comparison algorithm, the email is sent to the owner for acknowledging the activity. If the match was not found, the picture is sent to the owner to take further action. The email has the attachment of the photo of the intruder captured in the camera with the date and time and the respective message. The final action of the servo motor happens when the door/shutter unlocks and locks itself; the pi camera again switches back to the off condition after a certain delay. The PIR sensor maintains the state of "ON" to detect the next obstacle.

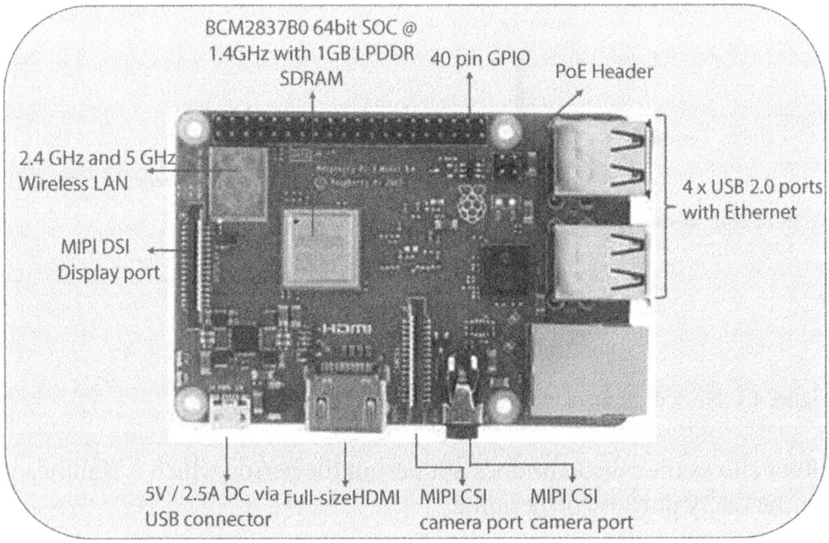

Figure 4.2 Raspberry Pi 3B+.

4.3.2 Hardware Requirements

4.3.2.1 Overview of Raspberry Pi Version 3 B+ Module

Raspberry Pi has a 1.4GHz processor, a dual-band 2.4GHz and 5GHz wireless computer network, Bluetooth 4.2/BLE, a faster LAN, and Edgar Allen Poe support via a special Edgar Allen Poe HAT. Standard compliance certification is included with the dual-band wireless computer network, allowing the board to be designed into finished products with far less wireless computer network compliance testing, increasing both the cost and time to market. The mechanical footprint of the Raspberry Pi Three Model B+ is comparable to that of the Raspberry Pi two Model B.

4.3.2.1.1 Features

The features available in raspberry pi are Broadcom BCM2837B0 and Cortex®-A53 (Arm®v8) 64-bit System on Circuit at 1.4 GHz. It has an LPDDR2 SDRAM of 1-gigabit of space. It also has a 2.4 rate along with a five rate IEEE 802.11.b/g/n/a wireless LAN, Bluetooth four.2, and BLE. Gigabit LAN over USB 2.0 (maximum output 300 Mbps) and an Extended 40-pin GPIO header with Full-size HDMI and four USB 2.0 ports are included. There is also a CSI camera port for attaching a raspberry pi camera, as well as a DSI show port for connecting a raspberry pi touch screen show. It also has a 4-pole stereo output and composite video interface, as well as a Micro memory card port for loading and storing information from an associated software package system. It features a DC power input of 5 V/2.5 A. Support for power over LAN (PoE) necessitates the purchase of an additional PoE HAT.

4.3.2.2 Overview of PI Camera

One of the lightest weight components among all other components in this whole design is the PI camera module. The main reason to use this module is that the raspberry pi 3 has a special port to accommodate this module, which makes the connection easier than any other camera module. For establishing a connection and sending and receiving signals this camera uses a protocol called an MIPI camera serial interface protocol and this has a wide range of applications. This module is very effective for police works says a survey. By the small size, it overtakes the traditional webcam used with personal computers these days.

4.3.2.3 PIR Sensor Overview

Human movement can be detected by a PIR sensor up to 10 metres away. This is a value of average, as the range of normal detection of humans is between 5m to 12m. PIR is made up of a pyroelectric sensor that can detect the amount of infrared emitted by an object. Several important projects and areas require monitoring of whether a person exited or entered the

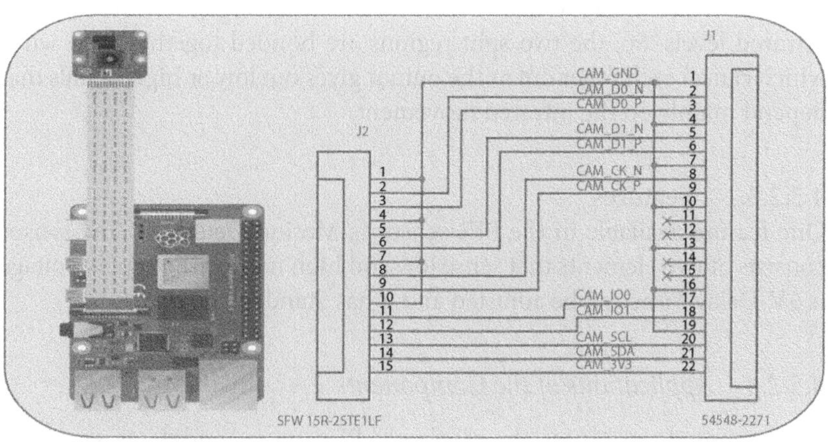

Figure 4.3 PI camera.

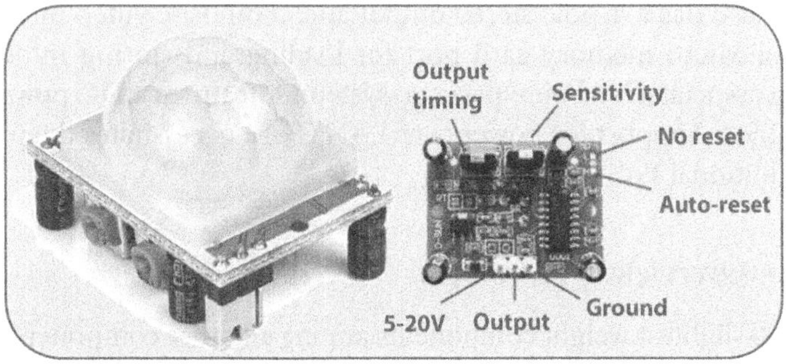

Figure 4.4 PIR sensor.

location. PIR sensors are great as they are flat controlled and of minimal effort. Their lens range is also wide and can be easily interfaced. The PI camera and PIR sensor with front as well as backside module is shown in Figures 4.3 and 4.4, respectively.

PIR sensors help in detecting motion but are also used to detect whether the person has left the range of the sensor. They are mostly used in home appliances and gadgets that are used for home or business purposes, mainly for their positive features such as being are smaller in size, consuming lower power, and being available at a lower cost. They are also simple to use. The PIR sensors are also called Passive Infrared, Pyroelectric, or IR motion sensors. The reason behind calling them that is that they can detect the various levels of infrared radiation. Every object gives out some low-level radiation but the hotter something radiates, the more energy it has. The sensor in a motion detector consists of two halves. That is mainly because we are trying to detect the movement (change), not the medium infrared levels. So, the two split regions are bonded together with wires which cancel each other out as the output gives out low or high signals that depend mainly on the infrared movement.

4.3.2.3.1 Features

One feature available in the PIR sensor is Motion Detection. The sensor consists of two elements that sense low and high noises. The supply voltage is 5V. Delay Time can be adjusted and it has Standard TTL Output.

4.3.2.4 Applications of the Components

The proposed work uses the raspberry PI camera module as shown in Figure 4.3 to observe live video spontaneously and detect the faces in it.

Raspberry pi camera works well with the raspberry pi module to connect effectively. The PIR sensor is used to detect the motion of the person or object present in the range. Once the obstacle is detected, the camera switches ON and starts monitoring.

4.3.3 Power Supply

In this project, the power is given to the raspberry pi's micro-USB port either through the computer or through the DC adapter. The DC adapter (power supply unit) is expected to deliver a power of 2.5A ideally, which is the recommended setting. But by connecting the micro-USB cable with the socket, it is ensured that the required current is transferred from the computer. So, a maximum of +5V is assumed sufficient for the working of the total circuit. Normally, any electrical equipment is powered with 220V and 50 Hz frequency. So, the PSU has a primary role of stepping down such a huge voltage into 12V and 50 Hz. For this purpose, two full-wave rectifiers are used to step the voltage down. Then, the output from the rectifiers may not be pure and might have some variations, so to rectify it the output signal is given to the filter to cut down the unwanted part. Then we have two ICs to regulate the voltage only to +5V or -5V if required. They are regulators LM7805 (for +5v) and LM7905 (for –5V). The block diagram of power supply is shown in Figure 4.5. It has following main components as

(i) Transformer
(ii) Bridge Rectifier
(iii) Filter
(iv) Regulator.

4.3.3.1 Transformer

The transformer is the device that is used to convert high AC (alternating current) into either high or low AC. They work on the principle of mutual

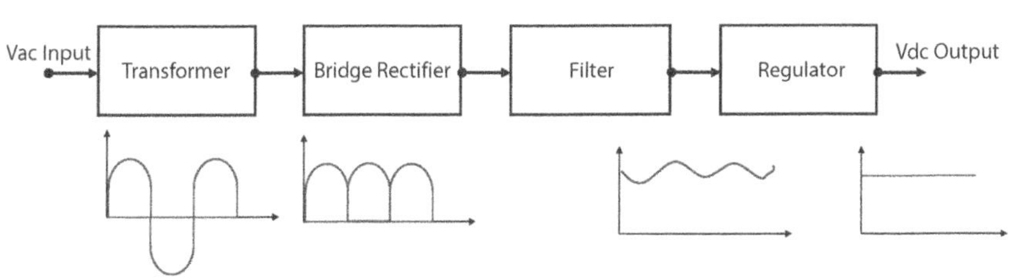

Figure 4.5 Block diagram of power supply.

inductance. The transformer has two kinds of windings. The first winding is called the primary and carries the current straight from the main, which is 220V/50Hz. Then, by the principle of mutual inductance, the current gets induced in the secondary windings. If the secondary windings are less in number compared to primary windings, it would suggest that it is stepping down the current and it is called a step-down transformer. If the secondary windings are found to be more compared to the primary, then it is stepping up the current and called a step-up transformer. There is one more type of transformer that has both the primary and secondary windings in equal numbers. This is called the matched transformer.

4.3.3.2 Rectifier

The rectifiers are placed next to the transformers in the PSU (Power Supply Unit). This is the important block that handles the conversion of the Alternating Current signal into a Direct Current signal. The rectifiers are mainly made up of semiconductor diode elements and depending upon the number of diodes in the circuit, there are 2 different classifications of rectifiers. One is called a half-wave and the other is called a full-wave rectifier.

4.3.3.3 Filter

A filter circuit is a device that consists of capacitors and is used to cancel out the AC part in the rectified output. The circuits help to figure out which part is the DC element and feeds it to the load. As mentioned earlier, there are various kinds of filter circuits available in the market, but the capacitance filter is the one that is being widely used and preferred. The main work of this capacitor is to reduce the harmonic or unusual bumps and spikes found in the rectified harmonic output.

4.4 Implementation of Proctored Face Lock System Using Python

In this automatic door opener system using facial recognition the easiest connection is the Pi Camera module because it can easily fit into the Raspberry Pi 3 dedicated port. A PIR sensor is next connected to General Pin I/O number 18 of the raspberry pi, after which an LED is connected as an indicator with a 1 Kilo OHM resistance at a GPIO pin 17 of the pi.

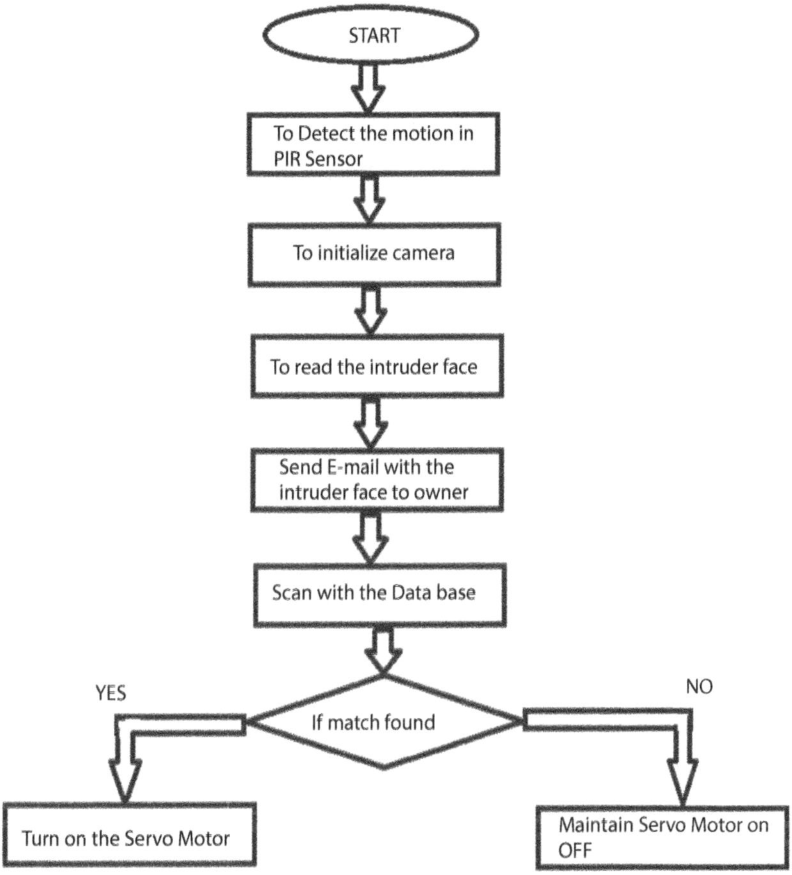

Figure 4.6 Flow chart proctored face lock system.

The above flow diagram shown in Figure 4.6 represents the implementation of the proctored face lock system using python. The database is the data in raspberry pi that is pre-trained with the face of the authorized person by using the Linear Discriminant Analysis algorithm.

4.5 Analysis and Discussion

Python is one of the languages with a large number of keywords. This makes coding easier and minimizes the number of lines in the program, which is why we choose the Python programming language. Before starting the coding process, the user needs to configure raspberry pi. After the

Figure 4.7 PIR sensor activates by detecting obstacles and in turn switches on camera.

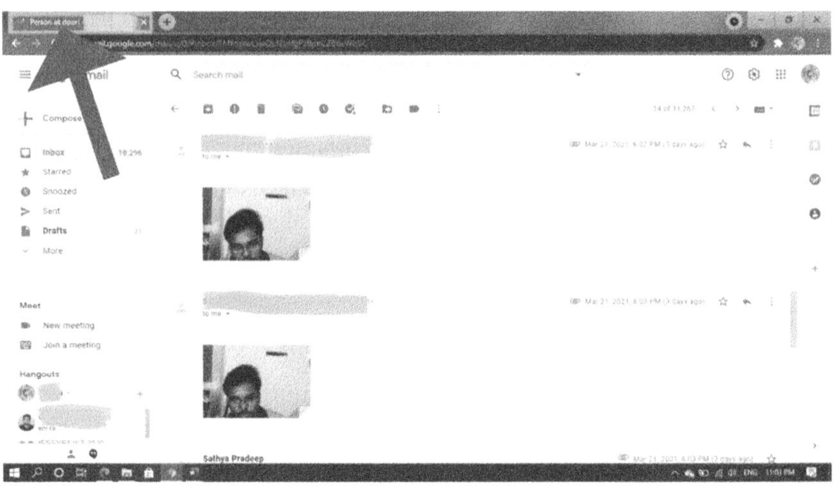

Figure 4.8 Intruder image shared with owner's mail.

Raspberry Pi OS and the Pi camera module have been successfully installed on the raspberry pi, we must install the pi camera library files in order to drive it with the raspberry pi. Prototype model of proctored face lock system is shown in Figure 4.7.

In such a manner before starting the main program, all the accessories such as the library for the PIR sensor must be installed. As the output message, called the intruder message as shown in Figure 4.8, is sent to the owner using mail, SMTP will be installed to do the process. Then, once the coding is completed, the hardware components can be assembled with the required kind of mechanical transmission to either the shutter or door.

4.6 Conclusion and Future Work

The proctored face lock system has been developed as a prototype model. It consists of a motor that rotates 90 degrees for a certain delay and then closes automatically by again rotating back 90 degrees in the opposite direction. Meanwhile, the mail would be sent with the photo of the intruder. If that algorithm is updated, soon it would make the device much more powerful. Here, the device is intended to be used for doors or shutters only, but by replacing the mechanical transmission part with other gadgets and making respective changes in the code we can use this device unconditionally with different circumstances and different parameters. The automatic door opener using a facial recognition circuit is cost-effective and power effective too. Also, the small size enables easy installation and guarantees that this lock mechanism will replace a lot of conventional types of locks. The data encryption and decryption give the most privacy and the data can be viewed only by the authorized person. Thus, the safest mode of authentication has been achieved.

References

Agarwal, R. and Kumar, P., (2019). A Face Recognition Method in the IoT for Security Appliances in Smart Homes, offices and Cities. *3rd International Conference on Computing Methodologies and Communication (ICCMC)*, pp.964-968.

Ahtsham, M., Yan, H. and Ali, U., (2019). IoT Based Door Lock Surveillance System Using Cryptographic Algorithms. *2019 IEEE 16th International Conference on Networking, Sensing and Control (ICNSC).*

Asra, S., (2019). Video Face Recognition using Deep Learning based Representation. *International Journal for Research in Applied Science and Engineering Technology*, 7(3), pp.53-58.

Begum, Q., (2019). GSM based Door Locking and Unlocking System. *International Journal for Research in Applied Science and Engineering Technology*, 7(9), pp.964-968.

Chitnis, S., Deshpande, N and Shaligram, A., (2016). An Investigative Study for Smart Home Security: Issues, Challenges and Countermeasures. *Wireless Sensor Network*, 08(04), pp.61-68.

Dong, T., Van, L. and Bao, P., (2021). Facial Expression Recognition Using Multi-Deep Convolutional Neural Network Encoders with Support Vector Machines. *International Journal of Machine Learning and Computing*, 11(5), pp.345-349.

F.A. Amuda and D.I.Tennyson, (2017). Design and Implementation of a Fingerprint Lock System. *IOSR Journal of Engineering*, 07(04), pp.13-19.

Fahim Faisal and Syed Akhter Hossain, (2020). Smart Security System Using Face Recognition on Raspberry Pi. *IEEE 2019 13th International Conference on Software, Knowledge, Information Management and Applications (SKIMA).*

Ghosh, D., (2021). Real- Time Attendance System using Face Recognition Technique. *International Journal of Engineering Applied Sciences and Technology*, 5(9).

Happy N. Monday, Jian P. Li and Grace U. Nneji, (2019). Design of an Improved Cost-Effective Electronic Locking System. *2018 IEEE 9th Annual Information Technology, Electronics and Mobile Communication Conference (IEMCON).*

Jana Alghamdi, Renad Alharthi, Raneem Alghamdi, Wafaa Alsubaie, Rawan Alsubaie and Dana Alqahtani, (2021). A Survey on Face Recognition Algorithms. *IEEE 2020 3rd International Conference on Computer Applications & Information Security (ICCAIS).*

Jiakailin Wang, Jinjin Zheng, Shiwu Zhang, Jijun He, Xiao Liang and Sui Feng, (2020). A Face Recognition System Based on Local Binary Patterns and Support Vector Machine for Home Security Service Robot. *IEEE 9th International Symposium on Computational Intelligence and Design.*

K. Lova Raju, V. Chandrani, SK. Shahina Begum and M. Pravallika Devi, (2019). Home Automation and Security System with Node MCU using Internet of Things. *International Conference on Vision Towards Emerging Trends in Communication and Networking (ViTECoN).*

Karimi, K., (2020). Secure Smart Door Lock System based on Arduino and Smartphone App. *Journal of Advanced Research in Dynamical and Control Systems*, 12(01-Special Issue), pp.407-414.

Kim, D. and Ban, C., (2015). The Design and Implementation of a Door-Lock System using a smart phone on Near Field Communication environments. *The Journal of the Korea institute of electronic communication sciences*, 10(11), pp.1217-1224.

N. Anusha, A. Darshan Sai and B. Srikar, (2018). Locker security system using facial recognition and One Time Password (OTP). *IEEE 2017 International Conference on Wireless Communications, Signal Processing and Networking (WiSPNET).*

N. Hema and Juli Yadav, (2020). Secure Home Entry Using Raspberry Pi with Notification via Telegram. *IEEE 2020 6th International Conference on Signal Processing and Communication (ICSC).*

Nag, A., Nikhilendra, J. and Kalmath, M., (2018). IOT Based Door Access Control Using Face Recognition. *2018 3rd International Conference for Convergence in Technology (I2CT).*

Prayag Bhatia, Shakti Rajput and Shashank Pathak, (2020). IOT based facial recognition system for home security using LBPH algorithm. *IEEE 2018 3rd International Conference on Inventive Computation Technologies (ICICT).*

R. S. Divya and Meera Mathew, (2017). Survey on various door lock access control mechanisms. *IEEE International Conference (ICCPCT).*

R.Manoj, Y.Rekha, R.Raju and A.Sharad, (2020). Smart Home Security System using IoT, Face Recognition and Raspberry Pi. *International Journal of Computer Applications*, 176(13), pp.45-47.

Sandesh Kulkarni, Minakshee Bagul, Akansha Dukare and Archana Gaikwad, (2017). Face Recognition System Using IoT. *IJARCET.*

Wang, L., Li, R., Wang, K. and Chen, J., (2014). Feature Representation for Facial Expression Recognition Based on FACS and LBP. *International Journal of Automation and Computing*, 11(5), pp.459-468.

Yadav, R., Agarwal, B. and James, S., (2018). Face Recognition System. *International Journal of Trend in Scientific Research and Development*, Volume-2(Issue-4), pp.1815-1818.

5

Advanced Adaptive Control of Nonlinear Plants

Santanu Mallick[1]* and Ujjwal Mondal[2]

[1]Department of Applied Electronics & Instrumentation Engineering, Bankura Unnayani Institute of Engineering, Bankura, India
[2]Instrumentation Engineering, Department of Applied Physics, University of Calcutta, Kolkata, India

Abstract
This chapter deals with advanced adaptive control applied to nonlinear plants. The U-Model control provides a general approach for modeling and controlling of different complicated systems. U-Model based control is simple and it is applicable to both linear and nonlinear plants. The objective of this current work is to apply MRAC using MIT rule, MRAC using Lyapunov stability method, MRAC augmented with PID method, Dynamic Inversion technique, U-Model based pole placement technique, and U-Model based MRAC technique with MIT rule for stabilization of a Single Inverted Pendulum on a cart. The MATLAB Simulink environment is utilized for simulation of different control strategies to determine the system performance.

***Keywords*:** Adaptive algorithm, dynamic inversion, feedback linearization, MRAC, nonlinear control systems, pole placement, U-model

5.1 Introduction

The single inverted pendulum is a popular example of a nonlinear system. The basic scheme of an inverted pendulum control system may be implemented in different industrial applications. It can be applied in

**Corresponding author*: santanu.mallick@gmail.com

Vipin Chandra Pal, Suman Lata Tripathi, and Souvik Ganguli (eds.) Industrial Control Systems, (77–112) © 2024 Scrivener Publishing LLC

different fields like the satellite launching application, robotics arm, transport machines which are required to balance objects, systems which support walking for patients, in robots which are used in domestic as well as industrial applications, and in object transport using drones. Basically, any system that requires vertical stabilization has dynamics that are similar to an inverted pendulum, for example in the human posture systems. Large scale constructions such as buildings are modelled as inverted pendulums. The Segway is also a popular commercial application of the inverted pendulum. Vibration control of a launching platform for shutters is also based on the concept of an inverted pendulum.

The system performance using a conventional controller is not suitable on account of small bandwidth, noise sensitivity, and dependability of error functioning to the plant due to change its parameters. In this situation, adaptive control is used wherein automatic adaptation of a controller is executed once the plant parameters of any plant are not identified and/or change with time [1–3]. In DI method coordinate conversion is implemented, just as if the given mathematical case arises linear in the transformed co-ordinate system. By applying the linear control design technique, the control action is implemented in that linear looking system. The controller is designed by enforcing a stable linear error dynamic [4]. The adaptive controller based on a recently developed U-Model based control has contributed to a generic approach for modeling and controlling various complicated systems [5].

R. Banerjee *et al.* has discussed the stabilization of a single inverted pendulum, using MRAC with an MIT rule by making different adaptation gains [6]. R. J. Pawar *et al.* has expressed the combination of MRAC accompanying a PID controller applied to an inverted pendulum to get better transient response [7]. The technique of MRAC along with a Linear Quadratic Regulator (LQR) for swinging up control of an inverted pendulum has been elaborated on by T. T. Sarkar *et al.* [8]. H. Fer *et al.* has illustrated the stability of a triple inverted pendulum placed upon a cart with the help of the DI technique [9]. R. Li *et al.* has explained a U-Model based design technique on different continuous time control systems [5].

5.2 Model Reference Adaptive Control

The MRAC structure has a reference model $G_m(s)$, plant $G_p(s)$, and controller along with an adaptive algorithm mechanism. An error signal $e(t)$ is developed by comparing the output of the reference model and that of the controlled plant. The adjustment mechanism structure determines exact

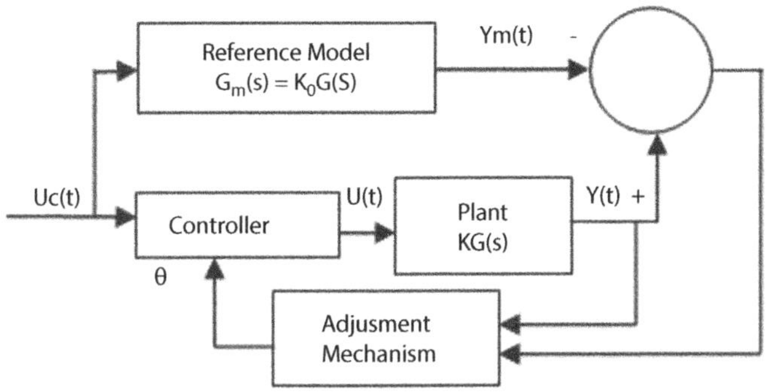

Figure 5.1 Scheme of MRAC.

control parameters employing different MRAC strategies and accompanied with these parameters, the controller structure generates the control input for the plant. The reference model is selected as per the desired system performance in such a way that the plant's output tracks the desired output. In this way, the controller enforces $e(t)$ towards zero (Figure 5.1). MRAC using the Lyapunov rule is used to ascertain the overall stability of a system [10–13].

Herein, the concerned plant has transfer function $KG(s)$, with K being an unknown parameter. The reference model has transfer function $K_0G(s)$, with K_0 being an unknown parameter. Also, $y(t)$ is the plant's output and $y_m(t)$ is the reference model's output. Therefore, the tracking error $e(t)$ becomes

$$e(t) = y(t) - y_m(t) \tag{5.1}$$

5.3 Dynamic Inversion

Dynamic Inversion (DI) is a technique by which it is possible to apply the available control for changing the differential equation in the desired manner. This technique was mainly developed for controlling the automatic flight control system in 1990 [14, 15].

Suppose the nonlinear differential equation of the plant is expressed as

$$\dot{x}(t) = f(x) + g(x)u \tag{5.2}$$

$$y(t) = h(x) \tag{5.3}$$

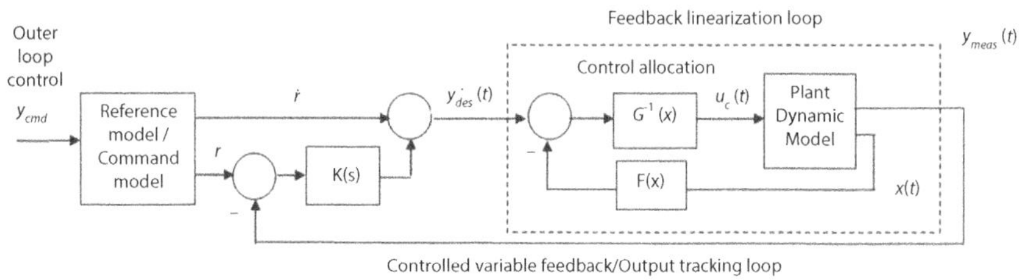

Figure 5.2 Scheme of dynamic inversion.

Here, $x(t)$ is the state vector and $y(t)$ is plant's output, i.e., measurement vector. $f(x)$ and $g(x)$ are time nonlinear vectors and $h(x)$ is a non-linear function. It is assumed that the system is linear for input vector u(t). $K(s)$ is the compensator.

Now, differentiating Equation (5.3) with respect to x, it is found that,

$$\dot{y} = \frac{\partial h}{\partial x}\frac{dx}{dt} = \frac{\partial h}{\partial x} f(x) + \frac{\partial h}{\partial x} g(x)u = F(x) + G(x)u \qquad (5.4)$$

In order to obtain that, $y(t)$ should chase the desired response and the signal due to the desired output dynamics $\dot{y}_{des}(t)$ has developed.

By inverting Equation (5.4), the control signal $u_c(t)$ is expressed as

$$u_c(t) = G^{-1}(x)[\dot{y}_{des}(t) - F(x)] \qquad (5.5)$$

The scheme of Dynamic Inversion technique is shown in Figure 5.2.

5.4 U-Model

The adaptive controller based on U-Model based control has established a fundamental approach for modeling and controlling of various complicated systems. U-Model deals with fewer parameters as well as variables, hence it provides a simple model configuration. This U-Model control technique is usually expressed as a polynomial or state space function and it is relevant for both linear and nonlinear dynamic systems. Being a polynomial structure, the U-Model adjusts the parameters of the system online in an adaptive way. The design procedure of a controller based on the U-Model method is straightforward. The U-Model method can be applied to most nonlinear discrete time dynamical systems, also it should be noted that the

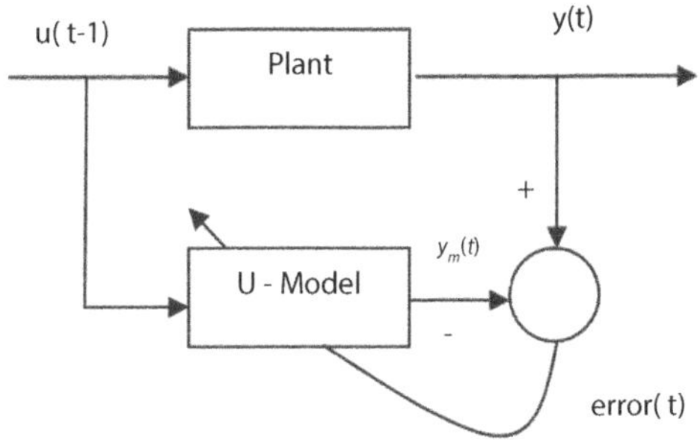

Figure 5.3 Scheme of U-model based design.

mapping is reversible. The basic scheme of a U-Model based design of a plant is shown in Figure 5.3. In this design, $u(t - 1)$ represents the current control input signal of the system, whereas $y(t)$ and $y_m(t)$ represent the output of the plant and plant model, respectively. The error(t) is the difference between $y(t)$ and $y_m(t)$ [5].

5.5 Single Inverted Pendulum

The dynamic of a single inverted pendulum system is presented in Figure 5.4. Here, the goal is to maintain for retaining the pendulum in the upright location, popularly known as swing control, hence the system is an unstable one. Under this circumstance, the pendulum is required not to fall, so a controlling force is required to retain the pendulum in the upright position. The desired objective is to provide the stable condition of the pendulum on the cart in a particular angle [6–8].

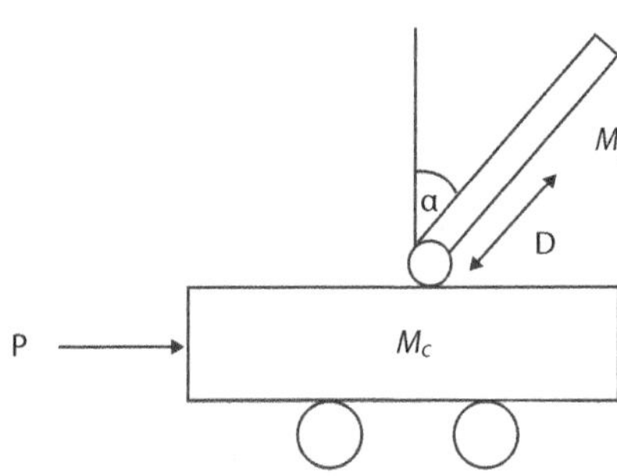

Figure 5.4 Simple inverted pendulum.

In case of a single inverted pendulum, it can be considered that the center of gravity (c.g.) of the rod of this pendulum is located at its geometric center and the rod is uniform.

Here,

M_C is the mass of cart
M_p is the mass of pendulum
D is the distance between c.g. of the pendulum and the pivot point
α being pendulum's angle
P represents force exerted to the cart

With the help of Taylor's series approximation, linearization can be done around the equilibrium point $\alpha = 0$.

Consider if α is small, then sin α tends to zero and cos α tends to unity.

Let the expression of the transfer function of the plant related to a single inverted pendulum can be considered as [6]:

$$G_P(s) = \frac{600}{s^2 + 8s + 600} \tag{5.6}$$

Here, a reference model is taken by choosing a standard second order differential equation having critical damping with natural frequency of oscillation of ω_n = 3rad/sec and damping ratio of ξ= 1 [7].

Consequently, the transfer function of the reference model can be expressed by:

$$G_m(s) = \frac{9}{s^2 + 6s + 9} \tag{5.7}$$

5.6 Performance Analysis

The performance analysis of the single inverted pendulum incorporating various techniques is interpreted in the following sections by applying a unit step input and considering the x axis as time (sec) and y axis as amplitude [16].

5.6.1 MRAC Employing MIT Rule

This rule had been revealed through MIT lab and in the first instance it was availed for outlining an automatic pilot project. Right now, this methodology may be executed for designing a compensator for all possible real-time applications.

In this methodology, cost function or loss function can be expressed in terms of error, i.e.,

$$J(\theta)=\frac{e^2}{2} \tag{5.8}$$

Here, θ represents the control parameter which is adjusted in order to minimize $J(\theta)$ [10, 12, 13].

In accordance with MIT rule, $\dot{\theta}$ is in direct proportion to the '–ve' gradient of $J(\theta)$, i.e.,

$$\frac{d\theta}{dt}=-\gamma\frac{\partial J}{\partial\theta}=-\gamma e\frac{\partial e}{\partial\theta} \tag{5.9}$$

where $\frac{\partial e}{\partial\theta}$ represents sensitivity derivative, which illustrates to what degree $e(t)$ depends upon θ. Hence, this rule relates the change of θ with $J(\theta)$. γ is adaptation gain, having a '+ve' value.

In case of $J(\theta) = |e|$, adjustment mechanisms will turn out as [11],

$$\frac{d\theta}{dt}=\gamma\frac{\partial e}{\partial\theta}\,sign(e) \tag{5.10}$$

This sign function/signum function extracts the sign of a real number, i.e., -1, 0, +1 for a negative number, zero number, and positive number, respectively. (5.11)

The desired objective in this technique is to construct a desired controller so that the plant model is capable to track the given reference model $G_m(s)$, here $G_m(s) = K_0G(s)$, and K_0 is a known parameter.

$$\text{Define the control law } U(t) = \theta U_c(t) \tag{5.12}$$

From Equation (5.1)

$$E(s) = KG(s)U(s) - G_m(s)U_c(s) = KG(s)\theta U_c(s) - K_0G(s)U_c(s) \tag{5.13}$$

By taking partial derivative,

$$\frac{\partial E(s)}{\partial \theta} = KG(s)U_c(s) = \frac{K}{K_0} Y_m(s) \tag{5.14}$$

From Equation (5.9) and Equation (5.14),

$$\frac{d\theta}{dt} = -\gamma \frac{\partial e}{\partial \theta} = -\gamma e \frac{K}{K_0} y_m \tag{5.15}$$

Equation (5.15) denotes law for adjustment of θ. The system is a stable one for a smaller value of γ [2, 12].

Applying the MIT rule and considering $\gamma = 0.1, 1, 2$, tracking characteristics of $G_p(s)$ including $G_m(s)$ are shown Figures 5.5–5.7.

From the tracking performance of the single inverted pendulum, it is realized that once γ is large, the model responds fast with a low value of oscillation and whenever γ is small, the model response is slow with a high value of oscillation, as shown in Figures 5.5–5.7. So, if γ increases, the accuracy of the single inverted pendulum increases as obtained by the data of performance indices as given in Table 5.1.

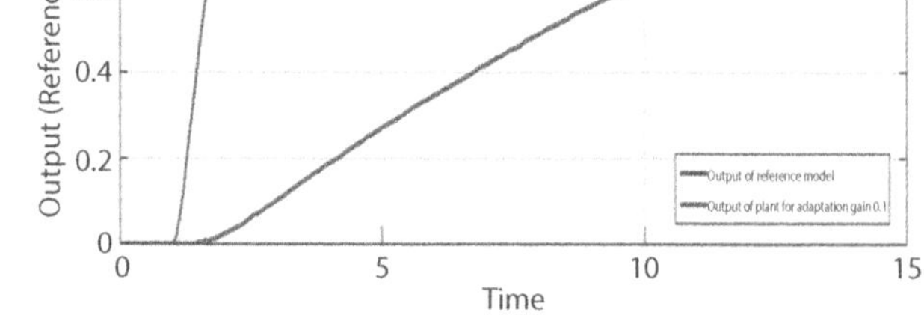

Figure 5.5 Tracking characteristics when γ = 0.1.

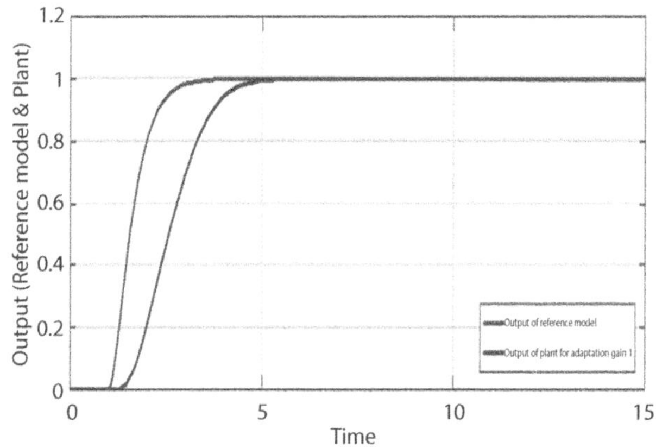

Figure 5.6 Tracking characteristics when $\gamma = 1$.

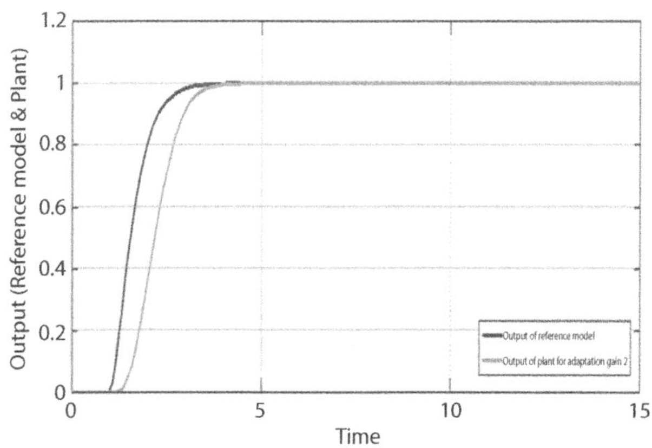

Figure 5.7 Tracking characteristics when $\gamma = 2$.

5.6.2 MRAC Employing LYAPUNOV Stability Method

In this method, a '+ve' definite Lyapunov function is assigned so that it's derivative in the expression turns '-ve' semi-definite, which makes the system stable. Whenever a derivative become '-ve' definite and at that instant the process becomes an asymptotic type of stable. Herein, adaptation methodology makes $e(t)$ towards zero [13, 17–19]. The scheme of MRAC employed with Lyapunov method is shown in Figure 5.8. Since the given plant is of second order, so $G_p(s)$ is analogous to following equation,

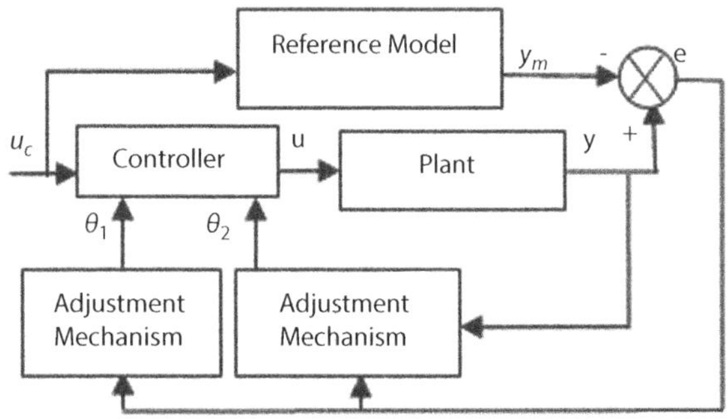

Figure 5.8 Scheme of MRAC employed with Lyapunov method.

$$\frac{d^2 y}{dt^2} = -A\frac{dy}{dt} + Bu \tag{5.16}$$

Considering the format of the plant as provided in Eq. (5.16), thereafter $G_m(s)$ is assumed as

$$\frac{d^2 y_m}{dt^2} = -A_m\frac{dy_m}{dt} + B_m u_c \tag{5.17}$$

In this case, y and y_m expresses the output of $G_p(s)$ and $G_m(s)$, respectively, while the reference input signal can be expressed as u_c. For the given method, the control law can be represented as

$$u = \theta_1 u_c - \theta_2\frac{dy}{dt} \tag{5.18}$$

In this case, θ_1 as well as θ_2 denotes control parameters and γ is gain which can be adjusted.

By Equations (5.1), (5.16), (5.17) & (5.18) the following may be obtained

$$\frac{de}{dt} = -A_m e - (B\theta_2 + A - A_m)y + (B\theta_1 - B_m)u_c \tag{5.19}$$

From the concept of error, it may be noticed that $e(t)$ becomes zero provided that,

$$\theta_1 = \frac{B_m}{B} \tag{5.20}$$

$$\theta_2 = \frac{A_m - A}{B} \tag{5.21}$$

A Lyapunov function may be expressed as [10, 18]

$$V(e,\theta_1,\theta_2) = \frac{1}{2}[e^2 + \frac{1}{B\gamma}(B\theta_2 + AS - A_m)^2 + \frac{1}{B\gamma}(B\theta_1 - B_m)]^2 \tag{5.22}$$

The given Lyapunov function becomes zero while e becomes zero. The controller's parameter is the same as the correct numerical. Moreover, $\frac{dV}{dt}$ ought to be negative [18, 19].

$$\frac{dV}{dt} = e\frac{de}{dt} + \frac{1}{\gamma}(B\theta_2 + A - A_m)\frac{d\theta_2}{dt} + \frac{1}{\gamma}(B\theta_1 - B_m)\frac{d\theta_1}{dt} \tag{5.23}$$

From Equation (5.19) and Equation (5.23)

$$\frac{dV}{dt} = -A_m e^2 + \frac{1}{\gamma}(B\theta_2 + A - A_m)\left(\frac{d\theta_2}{dt} - \gamma ye\right) + \frac{1}{\gamma}(B\theta_1 - B_m)(\frac{d\theta_1}{dt} + u_c e) \tag{5.24}$$

Therefore, the updated value of the parameters are given by

$$\theta_1 = -\frac{\gamma}{s}u_c e \tag{5.25}$$

$$\theta_2 = \frac{\gamma}{s}ye \tag{5.26}$$

For the given system, by employing $\gamma = 0.1, 1, 2$, the tracking characteristics of $G_p(s)$, including $G_m(s)$ are plotted in Figures 5.9–5.11. From the plots, it is noticed that when adaptation gain of the system increases, the characteristics of the system using the Lyapunov rule become fast. When γ increases, the accuracy of this pendulum system increases as pointed out by the data of various performance indices mentioned in Table 5.1.

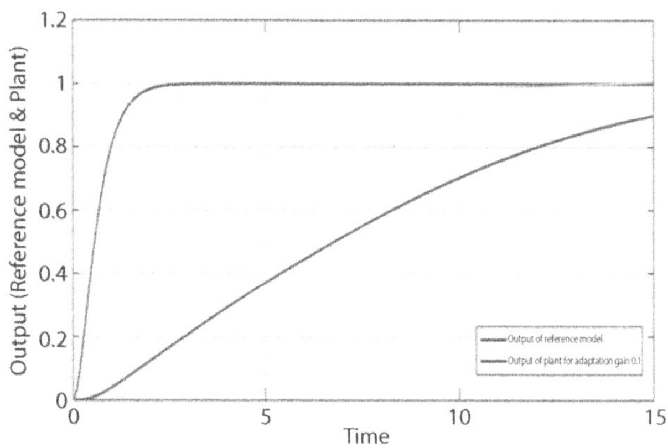

Figure 5.9 Tracking characteristics when $\gamma = 0.1$.

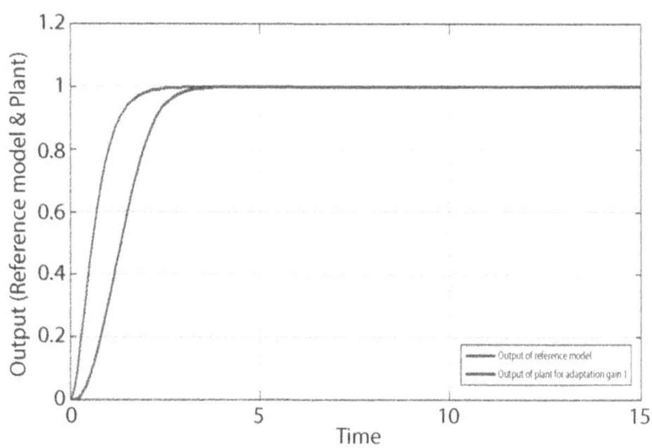

Figure 5.10 Tracking characteristics when $\gamma = 1$.

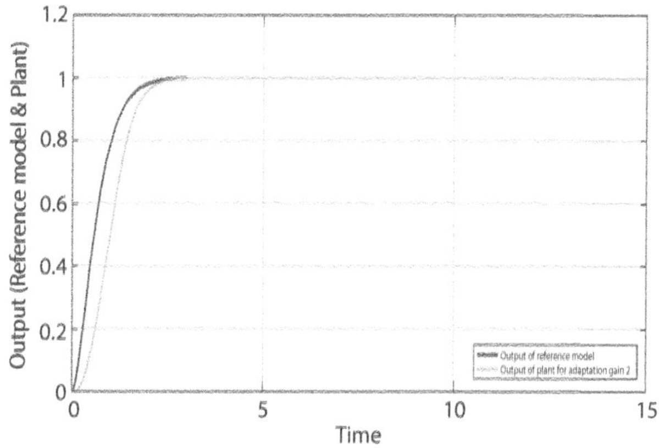

Figure 5.11 Tracking characteristics when $\gamma = 2$.

5.6.3 MRAC Augmented with PID Method

Considering a PID controller, $u(t)$ is stated as, [20]

$$u(t) = K_p e(t) + K_I \{ \int_0^t e(t) dt \} + K_D \frac{de(t)}{dt} \tag{5.27}$$

Herein, $u(t)$, K_p, K_i, and T_d mean control signal, proportional gain, integral gain, and derivative time, respectively.

In the case of modified PID algorithm methodology, the action due to the derivative section can be operated in a direct way upon feedback signal rather than $e(t)$ (Figure 5.12). Herein, a section of P action operates solely upon a reference value's portion and action due to D part is restored by the approximation which diminishes gain in increased frequencies. The modification of I action has been conducted so that it would never retain integrating whenever saturation occurred for the control variable [20–22].

In the case of a Modified PID control system, $u(t)$ is given by

$$u(t) = K_P e(t) + K_I \{ \int_0^t e(t)\, dt \} - K_D \frac{dy(t)}{dt} \tag{5.28}$$

The MRAC augmented with PID is an adaptive technique wherein the controller parameters are updated/tuned. Here, auto updating/tuning of K_p, K_I, *and* K_D are performed as per the control methodology based on

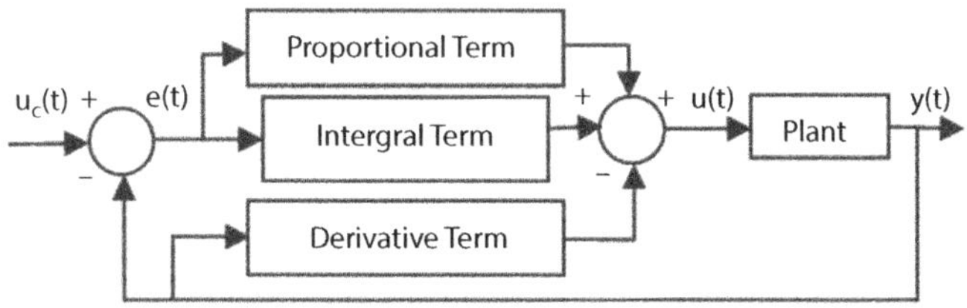

Figure 5.12 Scheme of modified PID control.

MRAC, as well as MIT rule, in order to ensure that the plant tracks concern the reference model, as shown in Figure 5.13 [22].

Assuming the second order plant is given as

$$G_P(s) = \frac{A}{s^2 + A_1 s + A_2} \tag{5.29}$$

By Equation (5.29), $u(s)$ is expressed as

$$u(s) = K_P\,[u_c(s) - y(s)] + \left[\frac{K_I[u_c(s) - y(s)]}{s}\right] - K_D y(s)s \tag{5.30}$$

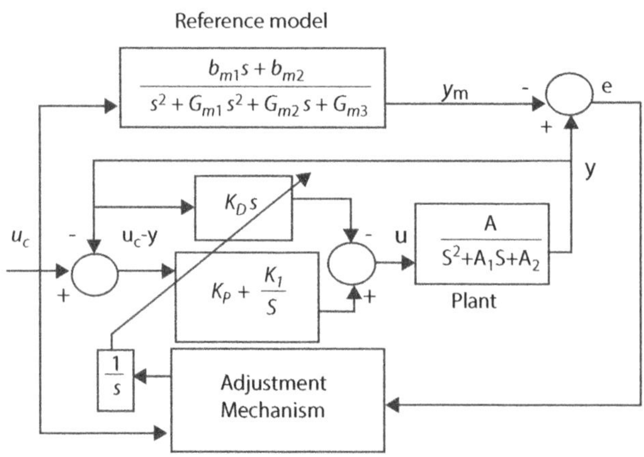

Figure 5.13 Scheme of MRAC with PID.

Moreover, by Figure 5.1, $G_P(s)=\dfrac{y(s)}{u(s)}$ (5.31)

As per Equation (5.30) and Equation (5.31),

$$\frac{y(s)}{u_c(s)}=\frac{A(K_P s+K_I)}{s^3+(A_1+AK_D)s^2+(A_2+AK_P)s+AK_I} \tag{5.32}$$

The PID controller is expressed as [22]

$$s^3+(A_1+AK_D)s^2+(A_2+AK_P)s+AK_I\approx s^3+a_{m1}s^2+a_{m2}s+a_{m3} \tag{5.33}$$

As per Equation (5.32) as well as Equation (5.33), the reference model will be

$$\frac{y_m(s)}{u_m(s)}=\frac{b_{m1.}s+b_{m2}}{s^3+a_{m1}.\ s^2+a_{m2}.\ s+a_{m3}} \tag{5.34}$$

By MIT rule, the updated values of the parameters of the PID system are

$$\frac{dK_P}{dt}=-\gamma_P\times\frac{\partial J}{\partial K_P}=-\gamma_P\times\frac{\partial J}{\partial e}\times\frac{\partial e}{\partial y}\times\frac{\partial y}{\partial K_P}=-\gamma_P.e.\frac{As.[u_c-y]}{s^3+(A_1+AK_D)s^2+(A_2+AK_P)s+AK_I} \tag{5.35}$$

$$\frac{dK_I}{dt}=-\gamma_I\times\frac{\partial J}{\partial K_I}=-\gamma_I\times\frac{\partial J}{\partial e}\times\frac{\partial e}{\partial y}\times\frac{\partial y}{\partial K_I}=-\gamma_I.e.\frac{A.[u_c-y]}{s^3+(A_1+AK_D)s^2+(A_2+AK_P)s+AK_I} \tag{5.36}$$

$$\frac{dK_D}{dt}=-\gamma_D\times\frac{\partial J}{\partial K_D}=-\gamma_D\times\frac{\partial J}{\partial e}\times\frac{\partial e}{\partial y}\times\frac{\partial y}{\partial K_D}=\gamma_D.e\frac{A.s^3[y]}{s^3+(A_1+AK_D)s^2+(A_2+AK_P)s+AK_I} \tag{5.37}$$

Equations (5.35–5.37) denote how different parameters of a PID controller change with respect to time.

Applying PID Tuning technique on a single inverted pendulum as mentioned in Equation (5.6), it is observed that K_P = 4.29570, K_I = 45.33646, and K_D = 0.101319.

Utilizing this data, the reference model can be expressed as

$$\frac{y_{m(s)}}{u_m(s)} = \frac{2577.42s + 27201.876}{s^3 + 68.7914s^2 + 3177.42s + 27201.876} \tag{5.38}$$

Taking $\gamma = 0.1, 1, 2$, and input as a unit step signal, it can be recognized that the proposed controller tends to track $G_m(s)$ and the performance is better for a large value of adaptation gain, as shown in Figures 5.14–5.16. In this situation, when γ increases, the accuracy of the given model increases as demonstrated by the value of performance indices, as shown in Table 5.1.

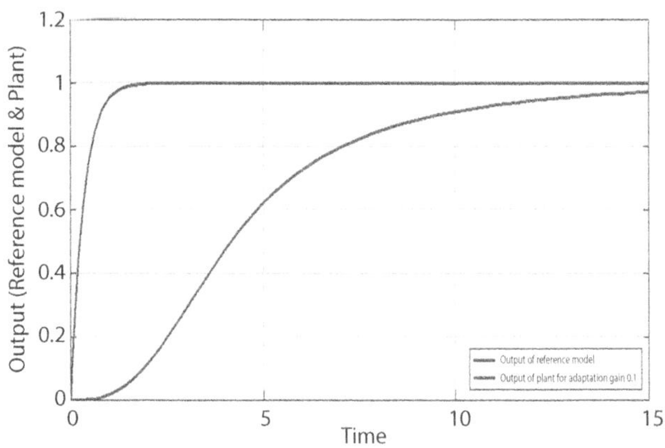

Figure 5.14 Tracking characteristics when γ =0.1.

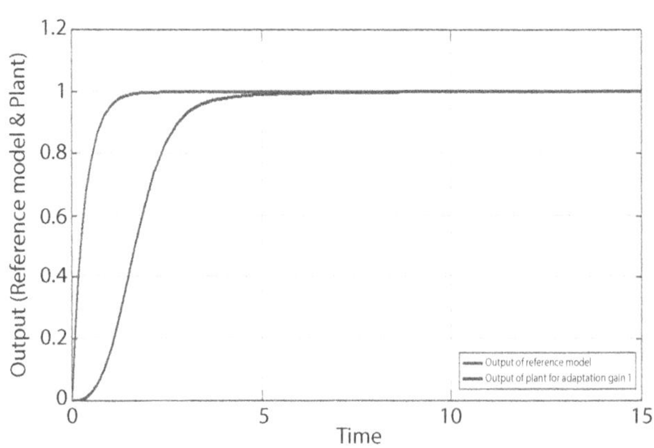

Figure 5.15 Tracking characteristics when γ =1.

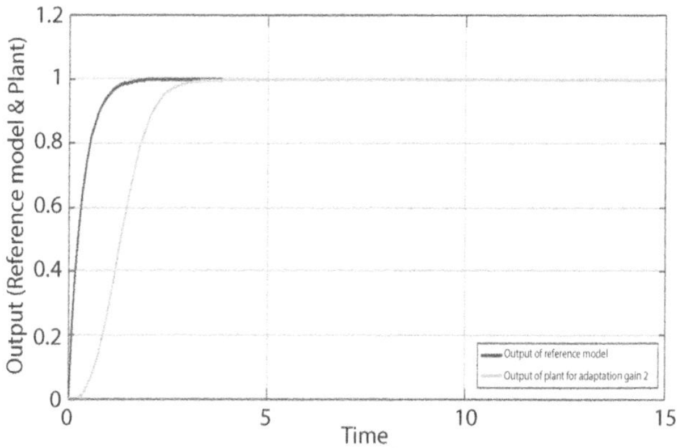

Figure 5.16 Tracking characteristics when γ =2.

In the course of adaptation time, PID parameters adjust themselves. The updated values of PID parameters may be determined graphically and their values will increase with an increase of γ, as shown in Table 5.2 [22].

5.6.4 Dynamic Inversion

Here, the performance analysis of a single inverted pendulum is elaborated with the Dynamic Inversion strategy. Here, the DI method can be implemented by feedback linearization which may be used in such a way that output of the nonlinear process is able to tracking a stable reference linear model.The scheme of MRAC using a feedback linearization technique has been elaborated in Figure 5.17. Here, a parameter estimator estimates the unknown time invariant parameters of a nonlinear control systems. The input state feedback linearization involving input transformation and state transformation, transforms a nonlinear system for the equivalent linear system. The adaptive controller reduces the error between an equivalent linear system and reference linear model output [4, 23–25].

Now, the plant as expressed by Equation (5.6) can be represented by the following differential equation:

$$\dot{x}_1 = -8x_1 - 600x_2 + 600u \tag{5.39}$$

$$\dot{x}_2 = x_1 \tag{5.40}$$

Table 5.1 Value of concern performance indices utilizing various adaptive control methodologies.

Technique used	Performance index	Adaptation gain		
		0.1	**1**	**2**
MRAC using MIT Rule	IAE	6.907	1.014	0.5917
	ISE	4.192	0.437	0.2091
	ITAE	44.54	2.381	1.188
	ITSE	22.39	0.9522	0.3981
MRAC using Lyapunov Method	IAE	6.691	0.7103	0.3561
	ISE	3.892	0.2513	0.0877
	ITAE	36.8	0.8336	0.311
	ITSE	16.55	0.2625	0.0684
MRAC Augmented with PID	IAE	4.712	1.45	1.021
	ISE	2.85	0.8453	0.5695
	ITAE	17.87	1.881	0.9759
	ITSE	7.032	0.8502	0.4599

Table 5.2 Updated values of parameters of PID.

PID parameters	Adaptation gain		
	0.1	**1**	**2**
K_P	0.0227	0.18	0.31
K_I	0.29	0.85	1.11
K_D	1.7	0.034	0.073

The reference model as mentioned in Equation (5.7) is interpreted by:

$$\dot{x}_1 = -6x_1 - 9x_2 + 9u \tag{5.41}$$

$$\dot{x}_2 = x_1 \tag{5.42}$$

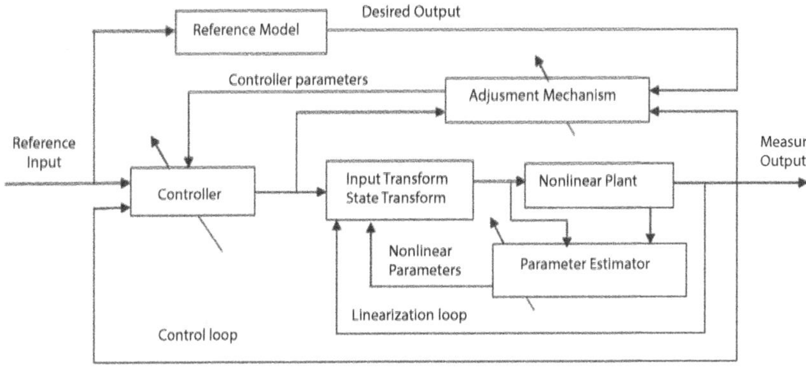

Figure 5.17 Scheme of MRAC employing feedback linearization.

The angle of pendulum α is represented as y, while y is same as the output of the system x_2.

In this situation, it is supposed that the inverted pendulum should not tilt beyond a certain range. It is assumed that this angle of tilt is 0.05 radian.

In case of the feedback linearization technique of a nonlinear system, regulation is necessary near the equilibrium point [4].

In mathematical statement, this will be

$$[x_1, x_2]^T \rightarrow 0, \text{ for } t \rightarrow \infty \tag{5.43}$$

Now, since $\dot{x}_1$ is nonlinear, it can be expressed as

$$\dot{x}_1 = \mathrm{F}_1(\mathrm{x}) + k + \mathrm{m}u \tag{5.44}$$

$$\text{Here, } \mathrm{x} \triangleq [x_1 \; x_2]^T \tag{5.45}$$

k and $\mathrm{F}_1(\mathrm{x})$ are the two constants.

Since $\dot{x}_2 = x_1$, linearization of $\dot{x}_2$ is not required.

Now, the controller is defined as u and it is expressed as

$$u(t) = u_0(t) + u_1(t) \tag{5.46}$$

Equation (5.44) can be written in a modified way as

$$\dot{x}_1 = \mathrm{F}_1(\mathrm{x}) - A\mathrm{x} + k + A\mathrm{x} + \mathrm{m}(u_0 + u_1) \tag{5.47}$$

$$\text{Here, } A = [a_{11} \;\; a_{12}] \triangleq \left[\frac{\delta F_1}{\delta X}\right]_{x=0} \tag{5.48}$$

Here, Ax is linear for the corresponding term x_1.

In this situation it is required for designing $u_0(t)$ so that nonlinear dynamics are cancelled out.

$$\text{Hence, } u_0(t) = -\frac{1}{m}[\mathrm{F}_1(\mathrm{x}) - Ax + k] \tag{5.49}$$

By Equation (5.47) and Equation (5.49), it is found that

$$\dot{x}_1 = \dot{A}x + mu_1 \tag{5.50}$$

Therefore, Equation (5.40) and Equation (5.50) represent a linear dynamic system.

In the technique from Equation (5.43), it should be mentioned that since $[x_1 \; x_2]^T$ tends towards zero as time tends towards infinity, it can be written as:

$$[\dot{x}_2 \;\; x_2]^T \rightarrow 0, \text{for } t \rightarrow \infty \tag{5.51}$$

Now the objective is to construct $u_1(t)$ in such a technique that the following equation, given as Equation (5.52), is validated

$$\dot{x}_2 + 2\xi w_0 \dot{x}_2 + w_0{}^2 x_2 = 0 \tag{5.52}$$

Here, ξ is a damping ratio and w_0 is natural frequency of oscillation.

By Equation (5.40), Equation (5.50), and Equation (5.52), it is found that

$$u_1(t) = -\frac{1}{m}[A\mathrm{x} + 2\xi w_0 x_1 - w_0{}^2 x_2] \tag{5.53}$$

From Equation (5.46), Equation (5.49), and Equation (5.53), it is found that

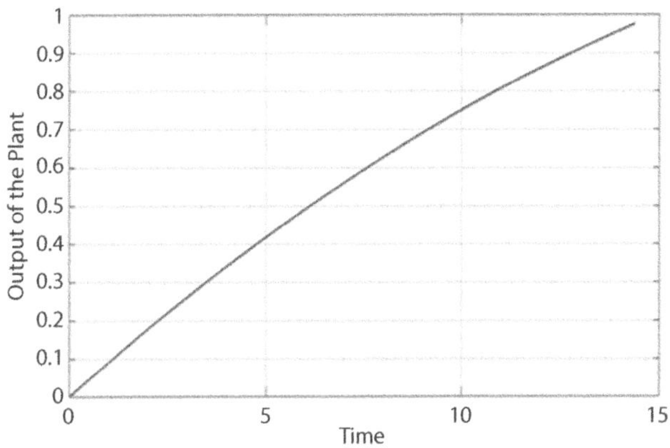

Figure 5.18 Response using dynamic inversion.

$$u(t) = u_0(t) + u_1(t) = -\frac{1}{m}[F_1(x) + k + 2\xi w_0 x_1 - w_0{}^2 x_2] \quad (5.54)$$

Hence, the solution of $u(t)$ validates $[\dot{x}_2\ x_2]^T \rightarrow 0$ for $t \rightarrow \infty$, hence $[x_1\ x_2]^T \rightarrow 0$ for $t \rightarrow \infty$. Therefore, it can be mentioned that the controller in this state of feedback methodology provides stability. The controller design technique used here is similar to the DI method. By using the DI method, it can be observed that the response of the given plant is sluggish as indicated by Figure 5.18.

5.6.5 U-Model Design Technique

The U-Model structure is basically a polynomial structure which consists of the parameters which are time varying in nature. This model is very useful for designing different ranges of nonlinear systems.

For a Single Input Single Output (SISO) U-Model system output of the plant can be represented as

$$y_m(t) = \sum_{i=0}^{M} \alpha^i (t-1) \quad (5.55)$$

Here, M denotes the degree of the polynomial and α_i represents the time-varying parameters which are past i/ps and o/ps except $u(t-1)$.

Here, the parameter set is $\alpha_i(t) = [\alpha_1, \alpha_2, \alpha_3, \ldots\ldots, \alpha_M]$ (5.56)

When applying the Normalized Least Mean Square principle, system parameters will adjust adaptively. The update model may be illustrated as

$$\alpha_i(t+1) = \alpha_i(t) + \mu(t)e(t)u^i(t) \tag{5.57}$$

Here, $\mu(t)$ is the learning rate and its value is in the range of 0 to 1. $e(t)$ is the mismatch error.

In this situation, $y_m(t)$ excites the controller through feedback and $u(t-1)$ drives the plant and model to give $y_m(t)$.

Here, the i/p to controller $x(t)$ may be assigned as $y_m(t)$. By replacing $y_m(t)$ and from Equation (5.55), it is found after calculation that

$$x(t) - \sum_{i=0}^{M} \alpha_i u^i(t-1) = 0 \tag{5.58}$$

For a U-model, an initial condition may be chosen as a previous control i/p and the Recursive Newton-Raphson method is applied for determining the control signal to solve Equation (5.58).

$$u_{k+1}(t-1) = u_k(t-1) - \frac{y_m(t) - x(t)}{y'_m(t)} \tag{5.59}$$

k is known as the iteration index [26].

Before designing a U-Model based pole placement technique, a brief discussion about the classical pole placement technique has been elaborated in the following portion.

5.6.5.1 *Pole Placement Based Controller*

A SISO system can be stated as

$$A_0(q)y(t) = B_0(q)[y_d(t)\{u(t) + d(t)\}] \tag{5.60}$$

Herein, A_0, B_0 represents polynomials in the forward/translation shift operator q. Here, $y(t)$, $y_d(t)$, $u(t)$, and $d(t)$ represent plant o/p, desired o/p, control i/p, and disturbance/error, respectively. The degree of A is n and degree of B is $n - d_0$. d_0 is pole excess which denotes the integer portion of the ratio of time delay and sampling period.

Here, d_0 = (no. of poles - no. of zeros). It is represented as delay operator q^{-1}. This is done due to

$$\overset{*}{A_0}(q^{-1}) = q^{-n} A_0(q) \tag{5.61}$$

This model now expressed as

$$\overset{*}{A_0}(q^{-1})y(t) = \overset{*}{B_0}(q^{-1})[u(t-d_0)] + d(t-d_0) \tag{5.62}$$

$$\text{where } \overset{*}{A_0}(q^{-1}) = 1 + a_1 q^{-1} + \dots\dots\dots + a_n q^{-n} \tag{5.63}$$

$$\overset{*}{B_0}(q^{-1}) = 1 + b_1 q^{-1} + \dots\dots\dots + b_n q^{-n} \tag{5.64}$$

where $m = n - d_0$

In discrete time, design technique turns purely algebraic. So continuous systems are given as

$$A_0 y(t) = B_0[y_d(t) + d(t)] \tag{5.65}$$

assuming A_0 and B_0 are relatively prime and A_0 is monic as a coefficient of the highest power and A_0 is one.

The linear controller is expressed as

$$R_0 y_d(t) = T_0 r(t) - S_0 y(t) \tag{5.66}$$

where R_0, T_0, and S_0 are polynomials and $r(t)$ is the reference i/p.

Hence, the controller as described by Equation (5.66) will be

$$y_d(t) = \frac{T_0}{R_0} r(t) - \frac{S_0}{R_0} y(t) \tag{5.67}$$

The Pole placement based controller as described by Equation (5.67) is represented by Figure 5.19.

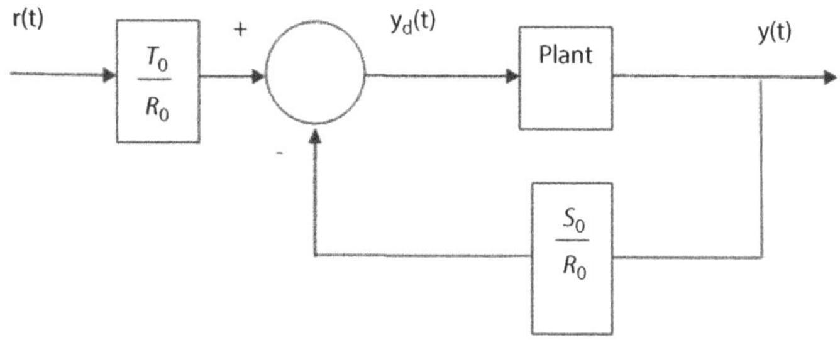

Figure 5.19 Pole placement based controller.

From Equation (5.65) and Equation (5.66)

$$y(t)=\frac{B_0T_0}{A_0R_0+B_0S_0}r(t)+\frac{B_0R_0}{A_0R_0+B_0S_0}d(t) \tag{5.68}$$

$$y_d(t)=\frac{A_0T_0}{A_0R_0+B_0S_0}r(t)+\frac{B_0S_0}{A_0R_0+B_0S_0}d(t) \tag{5.69}$$

Hence, the closed loop characteristics polynomial will be

$$A_0R_0 + B_0S_0 = A_{c0} \tag{5.70}$$

The Diophantine equation is used to solve Equation (5.70) and R_0 and S_0 can be determined. Now, it is required to determine T_0.

Now, the response of this dynamics system using command u_c can be expressed as

$$A_{0m}y_{0m}(t) = B_{0m}u_c(t) \tag{5.71}$$

From Equation (5.69), it is found that the following condition will be applicable as

$$\frac{B_0T_0}{A_0R_0+B_0S_0}=\frac{B_0T_0}{A_{c0}}=\frac{B_{m0}}{A_{m0}} \tag{5.72}$$

This condition as per Equation (5.72) indicates that factors of B_0T_0 & A_{c0} will be cancelled.

$$\text{The factor of } B_0 \text{ is } B_0 = B_0^{+}B_0^{-} \tag{5.73}$$

Here, B^+ is monic polynomial. If its zeros are stable and well damped, the controller may cancel these. So, B^- is being factor of B_{m0}.

$$\text{Therefore, } B_{m0} = B_0^{-}B'_{m0} \tag{5.74}$$

As B_0^- is cancelled out, it will be factor of A_{c0}. From the condition as per Equation (5.72), it is found that A_{mo} is also a factor of A_{c0}.

So, the characteristic polynomial for a closed loop system is

$$A_{c0} = A_x A_{m0} B_0^+ \tag{5.75}$$

As B_0^+ is a factor of B_0 & A_{c0} and from Equation (5.70), it is clear that it will divide R_0. Therefore,

$$R_0 = R'_0 B_0^+ \tag{5.76}$$

Therefore, the Diophantine expression will turn as

$$A_0 R'_0 + B_0^- S_0 = A_x A_{m0} = A'_{c0} \tag{5.77}$$

From Equations (5.73-5.76) it is found that

$$T_0 = A_x B'_{mo} \tag{5.78}$$

Suppose a discrete plant is described as

$$\frac{B_0}{A_0} = \frac{b_0 z + b_1}{z^2 + a_1 z + a_2} \tag{5.79}$$

Suppose the desired system is

$$A_{c0} = \frac{B_{m0}}{A_{m0}} = \frac{b_{m0} z + b_{m2}}{z^2 + a_{m1} z + a_{m2}} \tag{5.80}$$

Hence, the controller can be represented as

$$R_0 = z + \frac{b_1}{b_0} \tag{5.81}$$

$$S_0 = \frac{a_{m1} + a_1}{b_0} z + \frac{a_{m2} + a_2}{b_0} \tag{5.82}$$

$$T_0 = \frac{b_{mo}}{b_0} z \tag{5.83}$$

These calculations are used for controller design using the pole placement technique [27].

A classically based pole placement method is elaborated on in this section, considering the given single inverted pendulum as a plant.

As per Equation (5.6), the discrete transfer function of the plant with a sampling time of 1 sec can be described as

$$G_p(z) = \frac{0.9921z - .001256}{z^2 - 0.02081z + 0.0003355} \tag{5.84}$$

Also, as per Equation (5.7), the discrete transfer function of the reference model with a sampling time of 1 sec can be described as

$$G_m(z) = \frac{0.8009z + 0.1021}{z^2 - 0.09957z + 0.002479} \tag{5.85}$$

From the Equation (5.85), it is found that deg $A_0 = 2$, deg $B_0 = 1$.

The zeros and poles are located respectively at –0.12748158 and 0.049785 $\pm j0.000673628$

From Equation (5.79) and Equation (5.84), it is observed that

$b_0 = 0.9921$, $b_1 = -0.01256$, $a_1 = -0.02081$, $a_2 = 0.0003355$

From Equation (5.80) and Equation (5.85), it is observed that

$b_{m0} = 0.8009$, $bm_1 = 0.1021$, $am_1 = -0.9957$, $a_{m2} = 0.002479$

By Equation (5.81), Equation (5.82), and Equation (5.83), R_0, S_0, and T_0 can be determined as [27].

$$R_0 = z + \frac{b_1}{b_0} = z - 0.0126 \tag{5.86}$$

$$S_0 = \frac{a_{m1} + a_1}{b_0} z + \frac{a_{m2} + a_2}{b_0} \tag{5.87}$$

$$T_0 = \frac{b_{m0}}{b_0} z \tag{5.88}$$

By using Equations (5.86), (5.87), and (5.88), the values of $\frac{T_0}{R_0}$ and $\frac{S_0}{R_0}$ can be obtained, which is necessary to determine the desired controller [26–28]. The response of system and controller o/p by classical pole placement method are shown by Figure 5.20 and Figure 5.21, respectively.

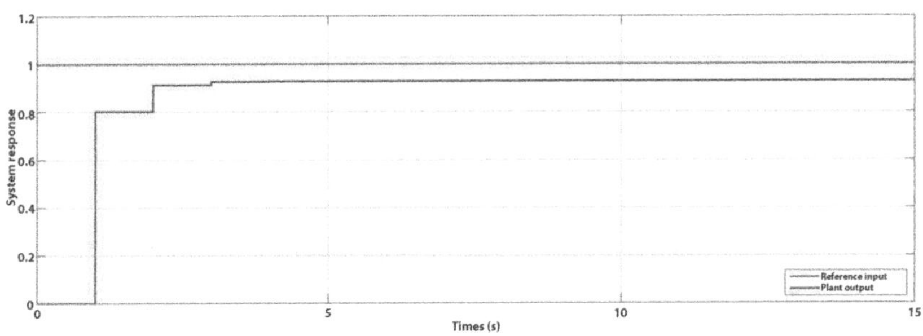

Figure 5.20 Response of system by classical pole placement method.

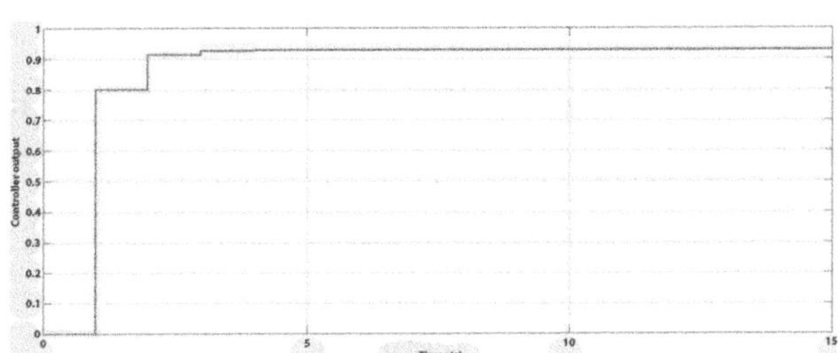

Figure 5.21 Response of controller o/p by classical pole placement method.

5.6.5.2 U-Model Based Pole Placement

For designing U-Model based pole placement methodology, it is required that the steady state error of a system be equal to zero.

$$\text{Now, } T_0 = A_{c0}(1) = 0.006779 \tag{5.89}$$

The polynomial R_0 and S_0 can be expressed as

$$R_0 = z^2 + r_1 z + r_2 \tag{5.90}$$

$$S_0 = s_0 z + s_1 \tag{5.91}$$

As per the Diophantine equation

$$R_0 = S_0 = A_{c0} \tag{5.92}$$

From Equation (5.85), Equation (5.90), Equation (5.91), and Equation (5.92), it is found that

$$r_2 + s_1 = 0.002479 \tag{5.93}$$

$$r_1 + s_0 = -0.9957 \tag{5.94}$$

To make convergence of $u(t)$ which is equivalently implied for making difference equation possessing stable dynamics, consider $r_1 = -0.9$ and $r_2 =$ 0.009

Hence, the characteristic equation of the sequence $u(t)$ can be expressed as

$$(z - 0.89)(z - .01) = 0 \tag{5.95}$$

The coefficient of polynomial S can be found with the help of the Diophantine Eq. as

$$s_0 = -0.0957 \tag{5.96}$$

$$s_1 = 0.006521 \tag{5.97}$$

The U-Model representation of the given plant as per Equation (5.6) can be expressed as [5]

$$y = \frac{\lambda_0}{s^2} + \lambda_1 u \tag{5.98}$$

$$\lambda_0 = -8\ \dot{y} - 600 \tag{5.99}$$

$$\lambda_1 = 600 \tag{5.100}$$

From Equations (5.66), (5.96), (5.97), (5.98), (5.99), and (5.100) and Figure 5.22, the response of the given system using the U-Model based pole placement method is shown in Figure 5.23, [27–32].

The output of concern controller of the given system using a U-Model based pole placement method is shown in Figure 5.24.

From the above discussion, it is clear that U-Model based pole placement design is concise as compared to classical based pole placement methods. In the case of the U-Model method, first specify the characteristic polynomial A_{c0}, then R_0 and S_0 will be calculated by Diophantine as given by Equation (5.92).

In classical based pole placement, as per Equation (5.70), $A_0 R_0 + B_0 S_0 = A_{c0}$. Here, A_0 and B_0 are the polynomial of the numerator and denominator of the plant model respectively that represent classical design based on the

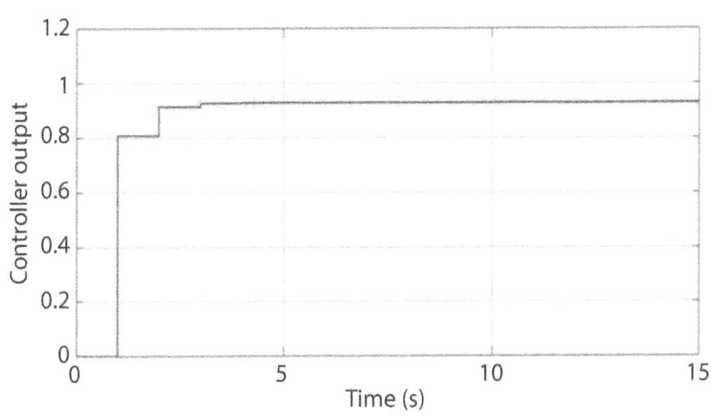

Figure 5.22 U-model based pole placement controller.

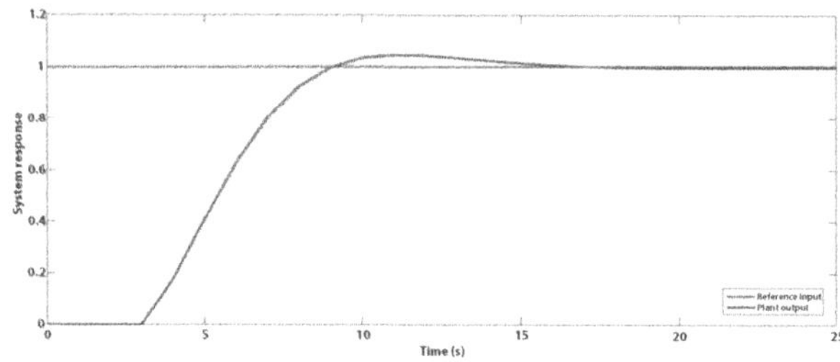

Figure 5.23 System response using U-model based pole placement method.

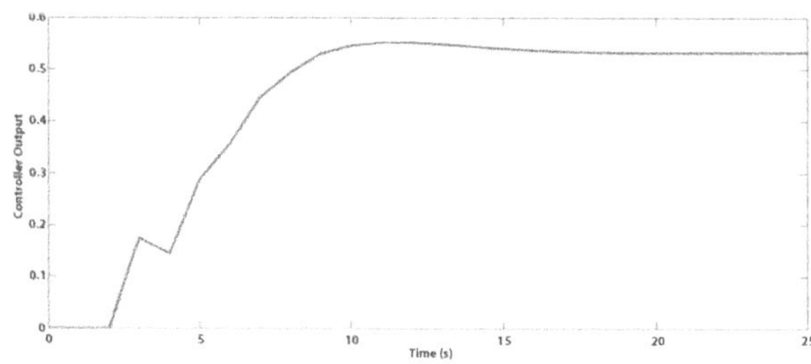

Figure 5.24 Output of controller using U-model based pole placement method.

plant model. When the plant model changes, this U-Model establishes the design law for R_0, S_0, and T_0.

The U-Model based pole placement technique will not rely on the model of the plant; here, the respective solution of the concerned controller o/p is involved in the plant model.

The U-Model design technique gives a basic approach to apply a linear approach for designing a control system associated with a nonlinear plant. The U-model design provides a good result in a linear control system. In classical U-Model design technique, the control design technique is involved for the plant model as well as controller for determining the controller o/p. In U-Model design technique design procedure is involved for designing a general controller and the next plant model is used for determining the o/p of the controller. In this way, the U-Model is better due to its conciseness and generality [27, 28, 31].

5.6.5.3 U-Model Based MRAC Technique with MIT Rule

For designing a U-Model based MRAC method, a nonlinear plant is represented as G. The root solver is represented as $\psi = U^{-1}$. The next reference model G_m and virtual plant model G_v are chosen. The reference i/p is y_r. The adaptive gain is denoted by γ. α is the coefficient and β is the constant. The o/p of the reference model and virtual o/p are denoted by y_m and y_v, respectively. The root solver is used to find out the root of the actual plant model. The o/p of the actual plant is y_p. The error is the difference between y_m and y_p. The U-model based MRAC scheme is shown in Figure 5.25.

Here, $\gamma = 0.1$, $\alpha = 0.01$, and constant $\beta = 2$. The reference i/p is chosen as a unit step signal. For the given plant, using a U-Model based MRAC technique based on the MIT rule and applying Equations (5.98–5.100), the response of the system is plotted in Figure 5.26. The output of the controller is provided by Figure 5.27, [31].

5.7 Conclusion

This paper illustrates the stabilization of a single inverted pendulum by applying MRAC employing the MIT rule, Lyapunov stability methodology, augmented with PID, feedback linearization based Dynamic Inversion technique, U-Model based pole placement technique, and U-Model based MRAC technique with MIT rule in a MATLAB environment. In the case of different MRAC methodologies, different adaptation gains are chosen

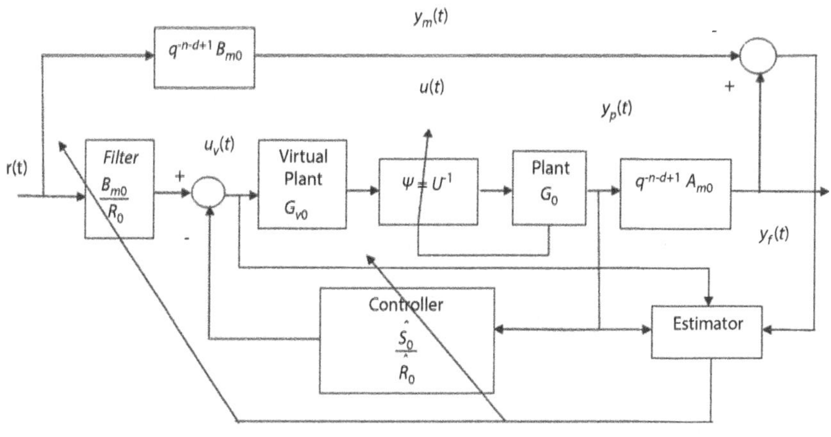

Figure 5.25 U-model based MRAC scheme.

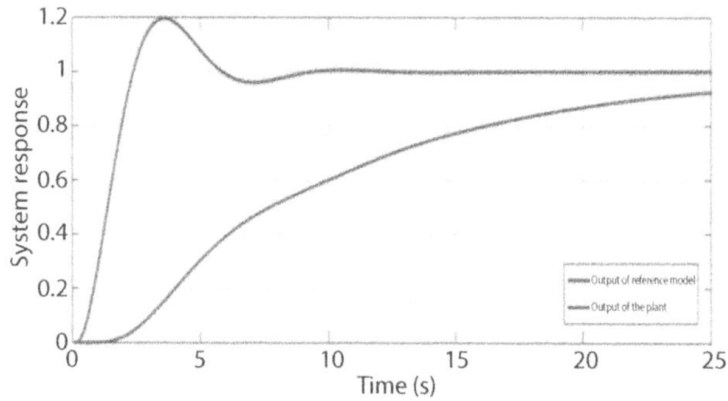

Figure 5.26 Response of system using U-model based MRAC technique with MIT rule.

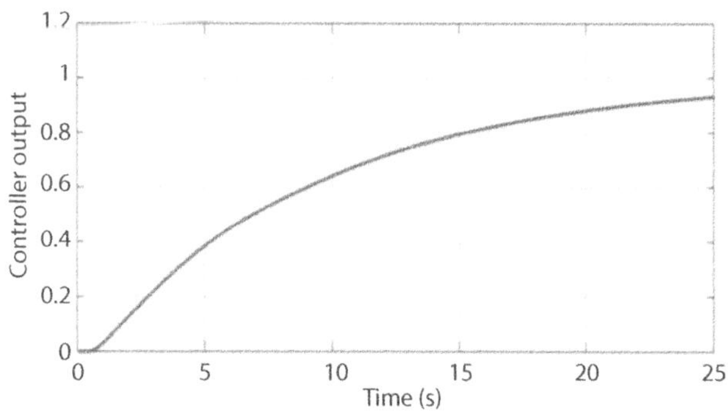

Figure 5.27 Output of controller using U-model based MRAC technique with MIT rule.

and a unit step input is applied in the system. Here, parameters of the controller are updated accordingly in order to achieve the desired closed loop system performance. In these MRAC methods, the different analysis shows that whenever adaptation gain progressively increases, then characteristics of the model possess quickly with low oscillation. From the plots as well as the value of different performance indices, it is observed that MRAC augmented with PID provides the supreme result among various MRAC methodologies which are elaborated in this paper. In the case of the MRAC augmented with PID technique, different PID parameters have been adjusted in an automatic way and these data have been evaluated. On the other hand, in a classical U-Model design technique, the control design technique is involved for plant models as well as controllers for determining the controller o/p. The U-Model design technique design procedure involves designing a general controller and a next plant model is used for determining the o/p of the controller. In this way, the U-Model is better due to its conciseness and generality. The U-Model based MRAC technique with MIT rule provides a slow response compared to the other techniques. From the response plots it is observed that MRAC augmented with PID method provides better result compare to the other techniques.

References

1. K. J. Astrom & B. Wittenmark. (2001). *Adaptive Control*, 2nd edn, 185-225.
2. P. Jain & M. J. Nigam. (2013). Design of a Model Reference Adaptive Controller Using Modified MIT Rule for a Second Order System. *Advance in Electronic and Electric Engineering*, 477-484.
3. P. Swarnkar, S. Jain & R. Nema. (2010). Effect of Adaptation Gain on system Performance for Model Reference Adaptive Control Scheme using MIT Rule. *World Academy of Science, Engineering and Technology*, 621-626.
4. R. Padhi. (2006). Feedback linearization based computer controlled medication design for automatic treatment of parturient paresis of cows. *Computer Methods and Programs in Biomedicine 84, Elsevier*, 19-26.
5. R. Li, Q. Zhu, J. Kiely & W. Zhang. (2020). Algorithms for U-Model-Based Dynamic Inversion (UM-Dynamic Inversion) for Continuous Time Control Systems. *Wiley*, 1-14.
6. R. Banerjee, N. Dey & U. Mondal. (2018). MRAC Based Approach for Stabilization of Single Link Inverted Pendulum. *IEEE International Conference on Current Trends toward Converging Technologies, Coimbatore*, 1-4.
7. R. J. Pawar & B.J. Parvat. (2015). Design and Implementation of MRAC and Modified MRAC technique for Inverted Pendulum. *IEEE International Conference on Pervasive Computing, Pune.*
8. T. T. Sarkar & L. Dewan. (2017). Application of LQR and MRAC for swing up control of Inverted Pendulum. *IEEE 4th International Conference on Power, Control & Embedded Systems, Allahabad.*
9. H. Fer & D. F.Enns. (1996). An Application of Dynamic Inversion to Stabilization of a Triple Inverted Pendulum on a Cart. *Proceedings of the IEEE International Conference on Control Applications*, 708-714.
10. A. Loria, E. Panteley & M. Maghenem. (2019). Strict Lyapunov functions for model reference adaptive control: application to Lagrangian systems. *IEEE Transactions on Automatic Control*, vol. 64, Issue 7, 3040-3045.
11. P. Swarnkar, S. Jain & R. K.Nema. (2011). Effect of Adaptation Gain in Model Reference Adaptive Controlled Second Order System. *ETASR Engineering*, 70-75.
12. J. Rothe, J. Zevering, M. Strohmeier & S. Montenegro. (2020). A Modified Model Reference Adaptive Controller (M-MRAC) Using an Updated MIT-Rule for the Altitude of a UAV, *Electronics*, 9(7), 1104,1-15.

13. P. Swarnkar, S. K. Jain & R. K. Nema. (2011). Comparative Analysis of MIT Rule and Lyapunov Rule in Model Reference Adaptive Control Scheme. *Innovative Systems Design and Engineering*, 154-162.
14. J. F. Horn. (2019). Non-Linear Dynamic Inversion Control Design for Rotorcraft. *MDPI Aerospace*, 1-25.
15. G. Ducard & H. P. Geering. (2008). Stability Analysis of a Dynamic Inversion Based Pitch Rate Controller for an Unmanned Aircraft. *IEEE RSJ International Conference on Intelligent Robots, France,* 360-366.
16. I. J. Nagrath & M. Gopal. (2009). *Control Systems Engineering*, New Age International Publishers, 5th edn, 223-224.
17. S. Mallick & U. Mondal. (2019). Comparative Performance Study of Lyapunov Based MRAC Technique and MRAC Augmented with PID Controller for Speed Control of a DC motor. *IEEE Second International Conference on Advanced Computational and Communication Paradigms (ICACCP), Gangtok.*
18. H. Tahersima, M. Saleh & A. Mesgarisohani, M.Tahersima. (2013). Design of Stable Model Reference Adaptive System via Lyapunov rule for Control of a Chemical Reactor. *Australian Control Conference,* 348-353.
19. M. Pal, G. Sarkar, R. K. Barai & T. Roy. (2017). Design of different reference model based model reference adaptive controller for inversed model non-minimum phase system. *Mathematical Modelling of Engineering Problems*, 75-79.
20. K. Pirabakaran & V M. Becerra. (2001). Automatic tuning of PID Controllers using Model Reference Adaptive Control Techniques. *Proceedings of the 27th Annual Conference of the IEEE Industrial Electronics Society*, 736-740.
21. A. Xiong & Y. Fan. (2007). Application of a PID Controller using MRAC Techniques for Control of the DC Electromotor Drive. *Proceedings of the 2007 IEEE International Conference on Mechatronics and Automation*, 2616-2621.
22. B. Singh & V. Kumar. (2015). A Real Time Application of Model Reference Adaptive PID Controller for Magnetic Levitation System. *IEEE Power, Communication and Information Technology Conference, Bhubaneswar.*
23. J. K. Lee, B. H. Suh & K.I. Abe. (1995). Model Reference Adaptive Control of Nonlinear System using Feedback Linearization. *SICE*, 1571-1576.
24. J. M. Biannic, L. Burlion & H. D. Plinval. (2014). Robust Control Design over Large Flight Envelopes:a Promising Approach for Aerial Robotics. *Journal Aerospace Lab*, 1-8.
25. R. Padhi & S. N. Balakrishnan. (2007). Optimal dynamic inversion control design for a class of nonlinear distributed parameter systems with continuous and discrete actuators. *IET Control Theory Application*,1, (6), 1662–1671.
26. E. Hasan, R. Ibrahim, S. A. A Taqvi & C. Olivier. (2017). Online System Modeling of Chemical Process Plant using U-Model, *IEEE 3rd International Symposium on Robotics and Manufacturing Automation.*

27. Q. Ji, Z. Ge, Z. Quanmin, N. Mokhtar & Y.Yufeng. (2016). Comparative Study on Pole Placement Controller Design with U-Model Approach & Classical Approach, *Proceedings of the 35th Chinese Control Conference, Chengdu,* 10436-10441.
28. Q. M. Zhu & L. Z. Guo. (2002). A pole placement controller for non-linear dynamic plants, Proceedings of the Institution of Mechanical Engineers, Part I: *Journal of Systems and Control Engineering*, 467-476.
29. W. Zhang, Q. Zhu, S. Mobayen, H.Yan, J. Qiu & P. Narayan. (2020). U-Model and U-Control Methodology for Nonlinear Dynamic Systems, *Wiley*, 1-13.
30. R. Li, Q. Zhu, P. Narayan, A.Yue, Y.Yao & M. Deng. (2021). U-Model-Based Two-Degree-of-Freedom Internal Model Control of Nonlinear Dynamic Systems, *Entropy*, 23(2), 169.
31. J. Qiu. (2019). Development of Adaptive control methodologies and algorithms for Nonlinear dynamic systems based on U-Control framework, *FET, Bristol,* 1-267.
32. Q. M. Zhu, D. Y. Zhao & J. Zhang. (2016). A general U-block model-based design procedure for nonlinear polynomial control systems, *International Journal of Systems Science,* vol. 47, No. 14, 3465–3475.

6

Design and Performance Analysis of Multiobjective Optimization Using PSO and SVM for PSS Tuning in SMIB System

R. Ramya[1], M.V. Suganyadevi[2]* and S. Usha[1]

[1]SRM Institute of Science and Technology, Kattankulathur, Chennai, India
[2]Saranathan College of Engineering, Trichy, India

Abstract

Energy networks are ever evolving largely due to the changes introduced into the production and loading system of dynamic behaviour and increased sustainability levels used by official organisation and also due to rapid use of wind energy, disseminated distribution, power warehouse systems, and so on. This creates a challenge for the design producers of energy systems. Today, most generators are installed on PSS. However, due to the complexity of the PSS, the correct adjustment of the PSS is crucial. The work focuses on the design and development of the integrated PSS and AVR action controller on the power system. The process of building an advanced control and the analysis presented in this work derives benefits from various modelling techniques. The purpose of the PSS design is to use Particle Swarm Optimization (PSO) and Support Vector Machines (SVM) to provide large lubrication torque without influencing the organising torque at analytic frequency vibration. The results of the use of the optimization algorithm in analysing the appropriate control for various operating conditions were analysed.

***Keywords*:** Synchronous generator, power system stabilizer (PSS), support vector machine, particle swarm optimization, real time simulation

**Corresponding author*: mvsuganyadevi@gmail.com

Vipin Chandra Pal, Suman Lata Tripathi, and Souvik Ganguli (eds.) Industrial Control Systems, (113–136) © 2024 Scrivener Publishing LLC

6.1 Introduction

Contemporary power systems are distinguished by greater system integration and increased reliance over the control for the efficient use of available resources. A reliable and economically viable energy supply is the key to industrial progress and the subsequent increase in quality of life. The increase in energy demand coupled with resource and environmental challenges poses a number of challenges for program planners. The demand for energy efficiency has been rising with the accessibility of electricity devices. Compatible generators play a crucial role for the power system stability [1]. The power supply level needs to match specific baseline conditions in terms of as presented: (a) frequency fluctuations, (b) power intensity, and (c) reliability level. To meet these needs, the production unit has extensive controller and entertainment controls [2].

In modern power systems, in addition to the generators present in immense quantity and compatible controllers, there are multiple kinds of loads, varying from light resistance to heavy loads and electronic controllers. The influx of multiple regulators and loads enhances the intricacy and inconsistency of the power system [3]. Because of this, the energy system is considered to be a complex system without a line that shows many problems of instability. Problems with instability in the electrical system that can lead to small or complete shutdown can lead to complications that can be generally divided as the following: 1) voltage, 2) angle, or 3) frequency. Although instability ultimately results in extinction or failure, by no means does it occur owing to sheer angle or voltage or frequency related problems and the first segment of the event may be certainly linked to one of the aforementioned stages [4].

Power systems are being developed as bounded load centres associated to a central system network. In the first phase such institutions were self-balanced and their amalgamation enabled them to increase power capacity thus assuring a standard reserve. This, although, drove to their joint fidelity and the power offset became very difficult. Thereafter, power outages occurred more than once. Many avalanches of the energy system arise due to frequency avalanches. Their principal cause was the power system integration, with links to a feeble and incomplete internal system [5]. Not in any way does this permit for the execution of administrative measures aimed at compliance with security regulations, i.e., (n-1) condition (setting of network shortcuts to maintain power to manage the power system in the event that any of the links is disjointed). Consequently, under emergencies the system was divided into sections, with one of them lacking in power. This has directed a profound recession and with a case as such, when the boiling pressure decreases the technical protection has impaired the producing units, thus leading to the elimination of their discharge [6].

The complication caused by Oscillations of Low Frequency and Stability Issues is because of system disruption. In an electrical network, plenty of interruptions that affect the power grid and affect the reliability of the generator are present. A few disruptions come from heavy loads. The generator will be affected by a sudden increase in power as considerable loads are abruptly connected to a power grid [7]. Loads, when abruptly loaded into the power system, can cause the generator frequency to tighten. It can also be contemplated as rotor angle oscillation of the generator. The oscillation can be instantaneously removed for a limited load, but the oscillation will persist longer before it is released when having an extensive load. The most unfavourable effect of an oscillation generator is not in sync and out of the generation safety transmission. Subsequent loss of the generator can lead to additional loads that abruptly affect other generators [8].

A few disruptions are induced due to errors. When more errors appear on the power grid, the safety transfer would hastily detect the error. If one time safety relays stop to respond to the respective error, it signifies that heavy load arises to the system. After that, this leads the generator of the rotor to rotate and disconnect it from the grid with the help of relays. In a few instances, because of the limit of transfer of power, this interference has a tendency be present in the grid until the minimum and maximum capacity of low loads. Power system stabilizers are developed to help reduce this impact by converting the generator miracle. As the power grid expands, the connection becomes increasingly complicated and further, because of the complex electrical systems of the non-linear system, is frequently given to low frequency fluctuation [9].

Today, most generators are installed on PSS. However, because of the complications of the PSS, the proper adjustment of the PSS is of utmost importance in the present day. Therefore, this concept is more focused on the preface of PSS and what method to tune it according to distinct control strategies and then to emulate real-time software with the help of dSPACE [10]. Shaft speed can be effortlessly detected by the electric power measurement. The stabilizer can deliver pure torque by utilizing a high filtration and low pass rate at a single electromechanical

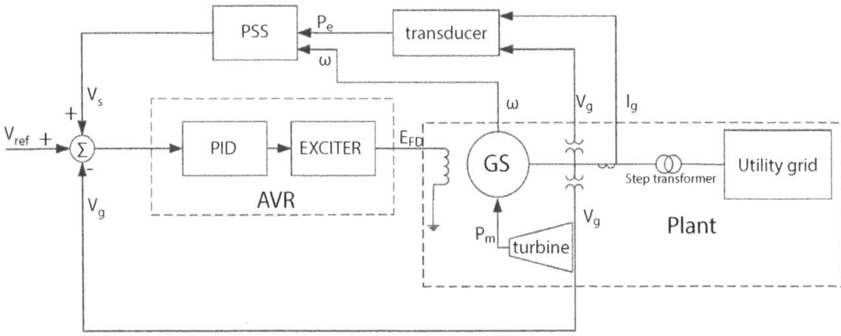

Figure 6.1 Power system feedback control.

frequency. The feedback power control system is depicted in Figure 6.1 [11].

The energy management scheme is schemed to attain two purposes. The former being control of the output power of the generator and the latter to enhance the rate of pull on the gas producer to boost the stability range of the closed system and to avert grid interference from the input from the producer [12]. This plant is composed of a power grid and a generator. A turbine that gives the necessary Pm helps to rotate the rotor generator. An EFD generated by AVR is the control signal in the plant which controls the Vg in the stator, i.e., the pout of the generator, which is obligated to rotate the rotor at synchronization speed. The power grid and stator are linked by a step transformer to contribute power to the grid [13, 14].

Figure 6.1 consists of a plant which has three effects and one input. The control signal is taken as the input that outputs AVR. Rotor speed rotation ω and terminal voltage V_g are the two effects to be controlled. The transducer uses the final voltage to obtain P_e power. To measure power change by virtue of rotor speed (ω) oscillation caused by fluctuation from the power grid, Pe and ω are used as PSS measurement signals. If there exists such an oscillation, the PSS is presumed to increment the pulse rate of the feedback system in order to diminish the fluctuation interruption from the power grid [15].

6.2 Small Signal Stability Analysis of SMIB System

Machines with equal power generation sites vibrate aversely to one another at 2.0 to 3.0 Hertz according to the unit classification and the reactance attaching them. This is called an intra plant, as the oscillation manifests itself inside a complicated structure. One generator switches the entire system from 1 to 2 Hz in local mode. The Optimization algorithms are nowadays extremely popular for designing in the area of engineering, especially due to the accessible and attainable rapid functioning computers. Effectiveness is elucidated as an art, procedure or method of doing something (such as drafting, plan or decision) as complete, efficacious, or ascetic as plausible. Numerous numerical optimization methodologies have been used to solve problems of accuracy.

The shortcomings of traditional methods are that they are trapped in the local optima and lead to untimely reunions. It stands in need of explicit mathematical functions and performance limits. Moreover, it creates complications when handling assorted variables. Random search procedures are implemented to vanquish this difficulty which comprises of evolutionary methods. It can be used for offline multimodal problems, usable performance issues and even plain sailing look-up space. Developmental calculations are a mathematical model for absolved utility and help to identify the right ideal global region swiftly. Such global optima are not accessible with a single computation but use mathematical steps to arrive at multiple separate calculations. An exploration process is formed after every computation as a good solution is reached. When an effort is made to develop a more prominent person, he or she engages in exploitation. Meta-heuristic expansion algorithms as swarm plots and evolutionary algorithms are mainly included by these methods. Evolution algorithms consist of genetic algorithms, evolutionary systems, evolutionary strategy, and genetic programs.

The most important statistics in the stability analysis of the power system are the rotating inertia calculations that explain the effect of the imbalance in joining the T_m and the T_e of each device.

The equations of acceleration are

$$\Delta\dot{\omega}_r = \frac{1}{2H}(T_m - T_e - K_D\Delta\omega_r) \tag{6.1}$$

$$\dot{\delta} = \omega_0\Delta\omega_r \tag{6.2}$$

Electrical Equations

$$\Delta T_e = K_1\Delta\delta + K_2\Delta\psi_{fd} \tag{6.3}$$

$$\Delta\psi_{fd} = \frac{K_3}{1 + pT_3}\left[\Delta E_{fd} - K_4 \Delta\delta\right] \tag{6.4}$$

$$\Delta E_t - K_5 \Delta\delta + K_6 \Delta\psi_{fd} \tag{6.5}$$

$$A\dot{v}_1 = \frac{K_5}{T_R}\Delta\delta + \frac{K_6}{T_R}\Delta\psi_{fd} - \frac{1}{T_R}\Delta v_1 \tag{6.6}$$

$$\begin{cases} \dot{x} = Ax + Bu \\ y = Cx + Du \end{cases} \tag{6.7}$$

$$\begin{bmatrix} \Delta\dot{\omega}_r \\ \Delta\dot{\delta} \\ \Delta\dot{\psi}_{fd} \\ \Delta\dot{v} \end{bmatrix} = \begin{bmatrix} \frac{-K_D}{2H} & \frac{-K_1}{2H} & \frac{-K_2}{2H} & 0 \\ \omega_0 & 0 & 0 & 0 \\ 0 & \frac{-\omega_0 R_{fd}}{L_{fd}} m_1 L'_{ads} & \frac{-\omega_0 R_{fd}}{L_{fd}}\left[1 - \frac{L'_{ads}}{L_{fd}} + m_1 L'_{ads}\right] & \frac{-\omega_0 R_{fd}}{L_{adu}} KA \\ 0 & \frac{K_5}{T_R} & \frac{K_6}{T_R} & \frac{-1}{T_R} \end{bmatrix} \begin{bmatrix} \Delta\omega_r \\ \Delta\delta \\ \Delta\psi_{fd} \\ \Delta v_1 \end{bmatrix} + \begin{bmatrix} \frac{1}{2H} & 0 \\ 0 & 0 \\ 0 & \frac{\omega_0 R_{fd}}{L_{adu}} \\ 0 & 0 \end{bmatrix} [\Delta T_m] \tag{6.8}$$

6.3 Real Time Simulation of SMIB

This segment analyses the construction of effective mechanisms for precise portrayal of synchronous machines for stability studies in energy systems. There are many mathematical models that can be utilized to measure the powerful performance of synchronization machines. The trade-off for a particular stability study of a model is between correctness and calculated price. These models may span from a simple classical imitation to complicated imitation.

6.3.1 dSPACE Simulated Flux Linkage Model of Synchronous Generator

A Synchronous Machine connection model was devised using MATLAB/Simulink. The model is made into a DSP (Digital Signal Processor) by making use of the Real Time Interface for dSPACE hardware, an actual-time simulation platform. The integration of the Synchronous Generator actual-time test platform has thus been enhanced to facilitate its control construction. The Simulink model of the SMIB system is created as depicted in Figure 6.2. Following the testing of replica in Simulink, it is required to devise for implementation on the actual time hardware. The I/O blocks replace the plant model from the Real Time Interface I/O library that fabricates

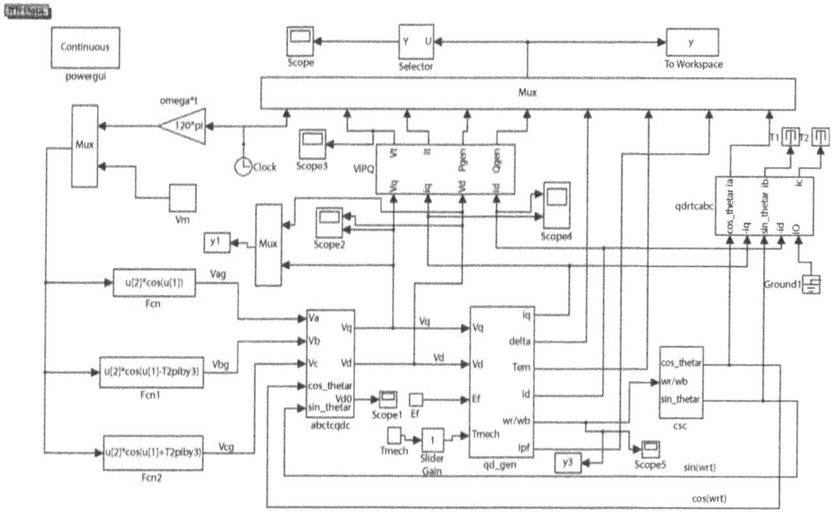

Figure 6.2 Single machine - infinite bus.

Table 6.1 Test machine parameters and system data.

Parameters of system	Values in p.u./sec
Winding resistance - Stator	0.0048
Synchronous reactance (d axis & q axis)	1.790, 1.660
Leakage reactance	0.215
Transient reactance (d axis & q axis)	0.355, 0.570
Sub-transient reactance (d axis & q axis)	0.275, 0.275
d-axis & q-axis transient time constant (open circuit)	7.900 sec, 0.410 sec
d-axis & q-axis sub-transient time constant (open circuit)	0.032 sec, 0.055 sec
Inertia constant, H	3.770

the interface to an actual controlled method. The test system taken is an 828 Mega Watts, 18Kilo Volt, 60 Hertz, 4 poles rated machine with the data displayed in Table 6.1 to analyse the performance of the model under actual time variations in excitation voltage and input mechanical torque. The temporary response of the model is further explored for the unanticipated three-phase short circuit fault generated in real time.

Following are the cases in which the machine's electrical performance is tested:

Case (i): Input mechanical torque changes
Case (ii): Excitation voltage changes
Case (iii): Fault in three phase short circuit

Case (i)
The T_{mech} variations in real time are provided using the dSPACE ControlDesk tool, as represented in Figure 6.3a. The input is suddenly reduced to zero from 1 per unit at t = 4 seconds. The sync machine is

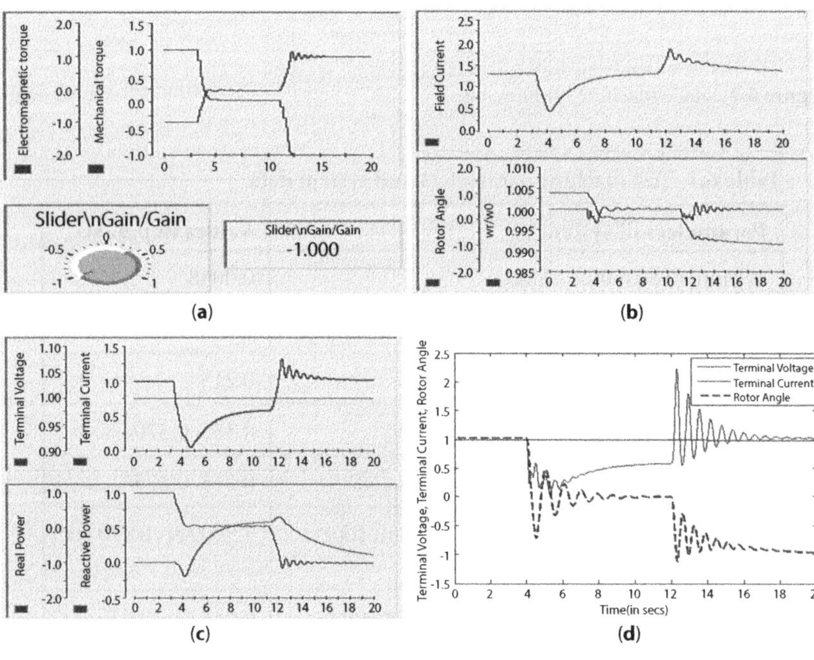

Figure 6.3 Response for changes in T_{mech}. (a) real time variation of Tmech; (b) rotor angle, field current, and speed; (c) P, Q, V, I; (d) terminal voltage and current, rotor angle.

authorized to operate as a motor with T_{mech} as -1 per unit at t = 12 sec. Figure 6.3a further shows the disparity of T_e for the changes in T_{mech}.

The variation of field current and rotor angle are illustrated in Figure 6.3b. In Figure 6.3c, the response of the machine for V_t, P, and Q recorded in the

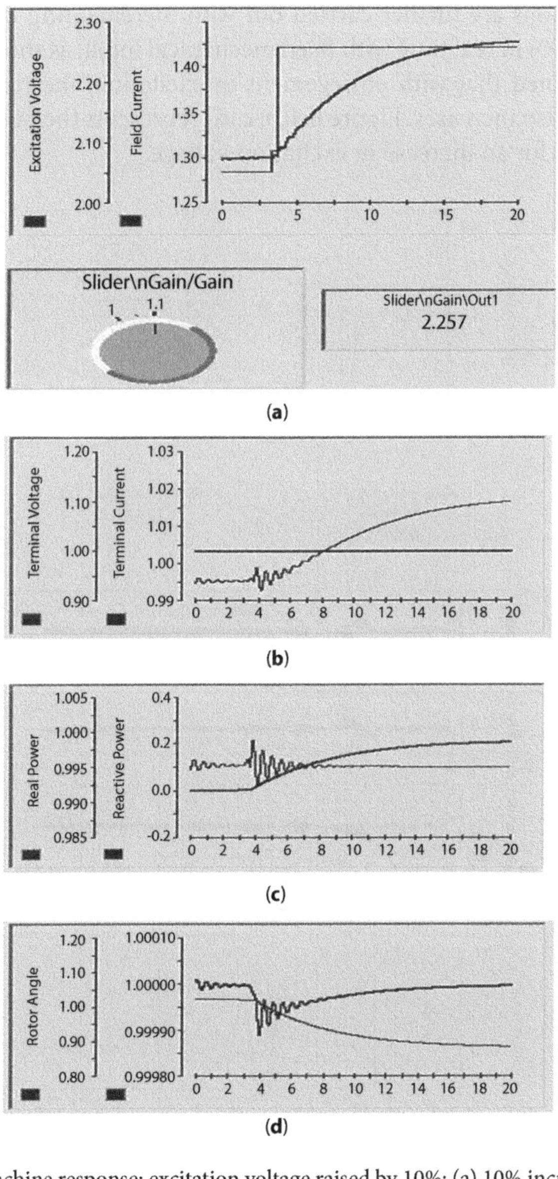

Figure 6.4 Machine response: excitation voltage raised by 10%: (a) 10% increase in excitation; (b) terminal current and terminal voltage response; (c) reactive power and real power response; (d) rotor angle and speed response.

real time with contrast of applied T_{mech} is shown.The stability of the synchronous machine is maintained even with the variation in mechanical torque. The same is carried out in offline simulation and presented in Figure 6.3d.

Case (ii)

The simulations are further carried out with incrementing the excitation voltage by 10% in real time with fixed mechanical input, as shown in Figure 6.4a. It is noted that with enlargement in excitation, the reactive power generation also increases. Figure 6.4(b, c, d) represents the performance of the machine for an increase in excitation voltage.

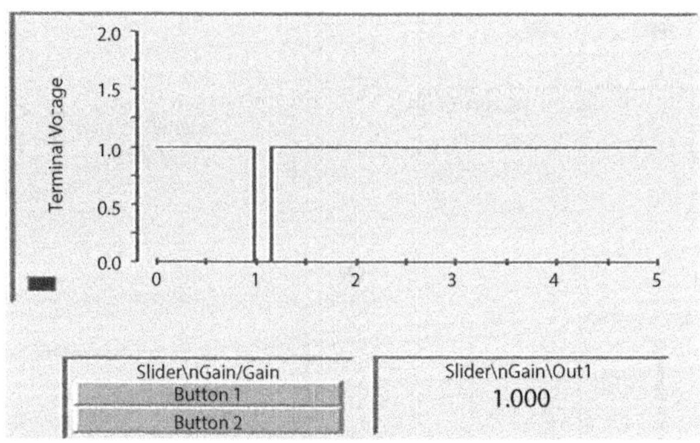

(a)

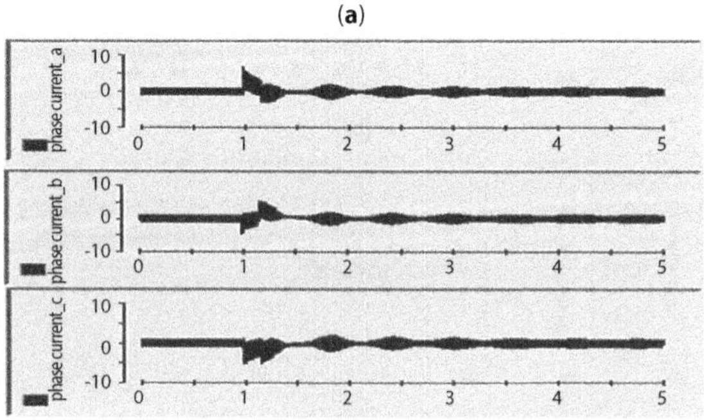

(b)

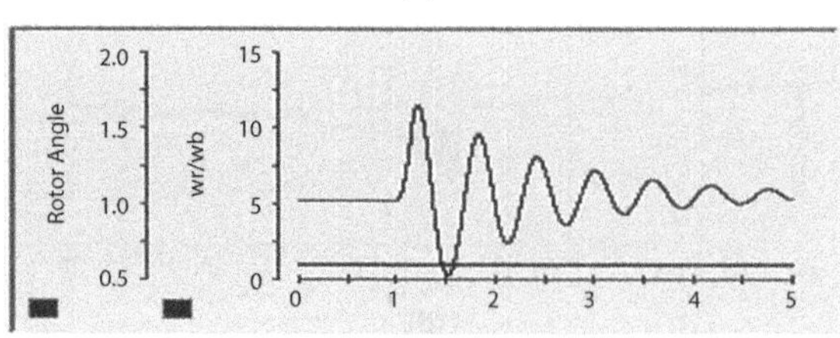

(c)

Figure 6.5 Three-phase short circuit for response of synchronous generator: (a) fault applied; (b) stator currents - three phase; (c) rotor angle and speed.

Case (iii)

In the present case, a three-phase symmetrical fault is generated at the terminals for about t=1sec and dismissed after 0.02sec, as in Figure 6.5a. Figure 6.5b shows the variation of three phase stator currents. Figure 6.5c depicts the response of rotor angle changes, indicating that the machine maintains stability after symmetrical fault is resolved.

6.3.2 dSPACE Simulation of Synchronous Generator's State Space Model (SSP)

The Synchronous Machine's state space model is created with the help of MATLAB/SIMULINK. The Simulink model for SMIB is depicted in Figure 6.6. The system is checked for its construction with the subsequent following in real-time.

Case (i):Variations in T_{mech}
Case (ii):Variations in V_{ref}

Case (i): The load position strength is kept persistent all around the program. The system is driven by the torque of the equipment used outside T_{mech}. $T_{mech's}$ sharp differences in the real time platform are provided using the dSPACE Desk Control Tool, as illustrated in Figure 6.7a. The machine input is raised from 1 per unit to 1.2 per unit at t = 50 sec. Figure 6.7a also shows the variation of T_{mech} and T_e. Terminal machine response E_t, P,Q, and other variables are displayed in this real-time deviation of the T_{mech}. Rotor speed, Rotor angle, and Terminal voltage are shown in Figure 6.7b.

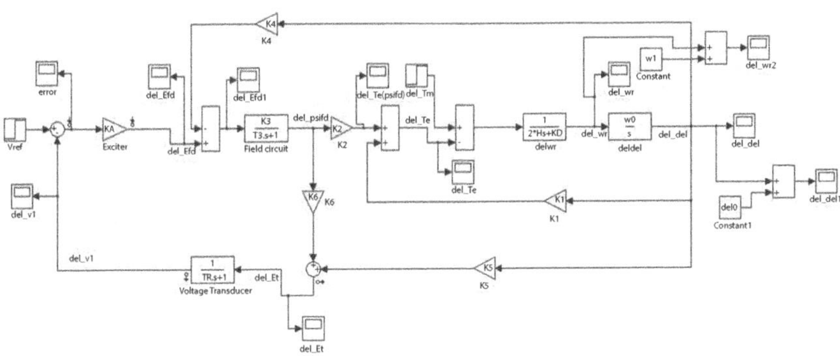

Figure 6.6 Synchronous generator model's simulation diagram.

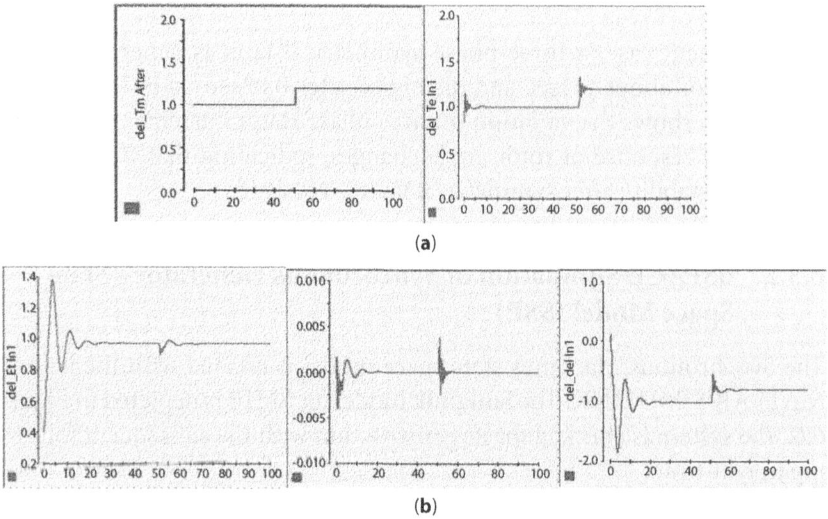

Figure 6.7 (a) Real time variations of electrical torque and mechanical torque. (b) Rotor speed and rotor angle, terminal voltage response.

Case (ii): With T_{mech} as a constant, a simulation is performed with a 20% reduction in reference values in real time. Figure 6.8a shows the decline in the reference voltage decreased to 0.8 per unit from 1 per unit.

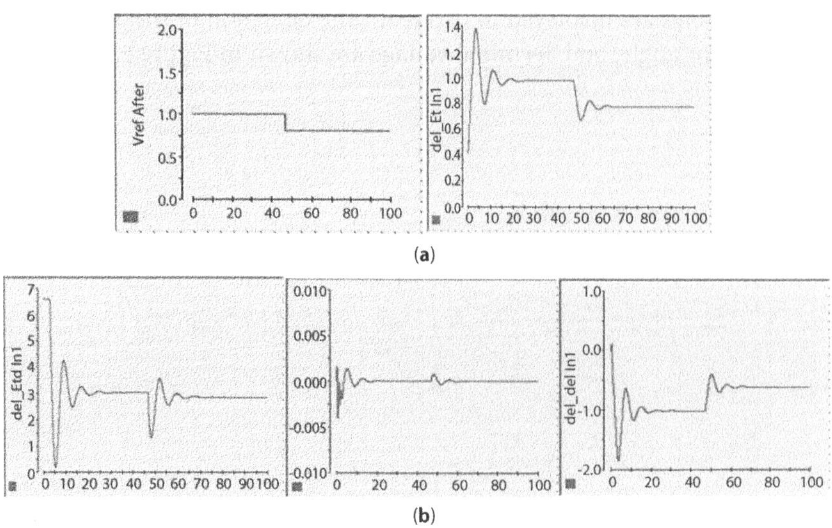

Figure 6.8 (a) V_{ref} and V_t variations. (b) Field voltage, rotor speed, and angle.

Figure 6.8a depicts the Terminal Voltage following the change in input reference (V_{ref}). Mechanical responses, like the V_f, deviation in ω, and Θ deviations for decrement in Vref are depicted in Figure 6.8b. This demonstrates that the machine achieves a stable state even though the perturbation takes place in real time.

6.4 Application of Optimization Techniques

6.4.1 Particle Swarm Optimization

Swarm intelligence is the characteristic of the system in which the integrated nature of local actors and their environment make global patterns work together to emerge. Swarm intelligence is a powerful way to harness power based on mobility and intelligence. Particle Swarm Optimization is a computer-aided design process that attempts to develop a solution for a person who will take into account a certain level of quality. The PSO solves the problem by employing more human solutions, referred to here as particles, and directing these particles into the search space using basic mathematical formulas in addition to particle structure and speed. However, each particle's movement is influenced by a well-known local position, as well as the most well-known locations in the search space that are revised as the best positions obtained by other particles. This is expected to move the crowd to more advanced solutions.

The PSO was originally said to have been created by Kennedy and Eberhart and was originally intended to mimic social behaviour, as in a flock of birds or a school of fish, an image created by the movement of living organisms. The algorithm was simplified and demonstrated to be effective. Many aspects of the PSO's philosophy and the multiplicity of spies are described in Kennedy and Eberhart's book.

6.4.2 Support Vector Machine Algorithm

The support vector regression design is depicted by the following design [16, 17].

The following is an example of a regression training set.

$$D = \{(x_1, y_1), (x_2, y_2).......(x_m, y_m\} \tag{6.9}$$

where $\mathbf{x}_i$ is an n-dimensional vector and y is the real number corresponding to each $\mathbf{x}_i$. The SVR function $F(\mathbf{x}_i)$ makes a mapping from an input vector $\mathbf{x}_i$ to the target y_i and takes form.

$$F(X) = w * x - b \tag{6.10}$$

$$\text{Minimize: } L(w) = \frac{1}{2}\|w\|^2 \tag{6.11}$$

$$\text{Subject to: } y_i - w * x_i - b \leq \varepsilon \tag{6.12}$$

$$w * x_i + b - y_i \leq \varepsilon \tag{6.13}$$

The optimization problem can then be rewritten as follows.

$$\text{Minimize: } L(w,\xi) = \frac{1}{2}\|w\|^2 + C\sum_i \left(\xi_{2i}, \hat{\xi}_{2i}\right), C > 0 \tag{6.14}$$

Subject to:

$$y_i - w * x_i - b \leq \varepsilon + \xi_i \qquad \forall (x_i, y_i) \in D \tag{6.15}$$

$$w * x_i + b - y_i \leq \varepsilon + \hat{\xi}_i \qquad \forall (x_i, y_i) \in D \tag{6.16}$$

$$\xi_i, \hat{\xi}_i \geq 0 \tag{6.17}$$

The constant $C > 0$ is the trade-off parameter between margin size and error amount.

$$\begin{aligned} &-\sum_i \alpha_i(\varepsilon + \eta_i - y_i + w.x_i + b) \\ &-\sum_i \hat{\alpha}_i\left(\varepsilon + \hat{\eta}_i + y_i - w.x_i - b\right) \end{aligned} \tag{6.18}$$

$$\hat{\alpha}_i \geq 0, \alpha_i \geq 0 \tag{6.19}$$

$$0 \leq \alpha, \hat{\alpha} \leq C \tag{6.20}$$

Finally, employing the kernel function, the SVR function F(x) becomes the following.

$$F(x) = \sum_i \left(\alpha_i - \hat{\alpha}_i\right) K x_i x_j + b \tag{6.21}$$

6.5 Real-Time Simulation of Single Machine System Using PSO-PSS

The purpose of the PSS design is to give additional toramp without influencing the synchronizing torque at critical oscillation frequencies. This can guide for an increment in settling time. This work therefore gives a central investigation on the large PSS value parameters through the use of Particle Swarm Optimization to minimize low oscillations, such as both pass and resolution times are decreased simultaneously, and the decrement of these components will significantly upgrade the stability of the power system. Therefore, the appropriate PSS parameter settings are one of the key issues in the control of the control system because precise parameters increase the system's robustness and some parameters can reduce the system's robust strength.

The failure circuit serves as a high pass filter and should transmit all interested waves. If only local routes are of interest, the continuous Tw period can be selected for grades 1 to 2. However, if the local roads are also to be dumped, then Tw should be selected from 10 to 20 degrees. There is noticeable improvement in the initial stability of the swing as Tw rises from 1.5 to 10. Tw's high value also enhances the final extreme power outcome during system island states.

$$T(S) = K_s (1+sT_1) / (1+sT_2) \tag{6.22}$$

Wherein Ks1 gains PSS and time zones, T_{11} to T_{22} have been selected to give the leading signal input range in the range of interest frames (0.1 to 3.0 Hertz).

1. For a specified period, the T_1 to T_2 in the equation should be selected from the compensation demand of the phase in order to obtain damping torque.
2. The benefit of the PSS should be chosen to give sufficient compliance with all critical procedures under many working states. It should be noted that the PSS is considered a specific working state (full load state with a strong or weak AC system) that is very important. While PSS can be used to provide complete flexibility under this scenario, the result may not be ideal under certain circumstances. Critical approaches include local and regional approaches and alternatives (called control or recreational modes) introduced with an exciter and/or torsional filter.

The optimization algorithm using ITE as the objective function is run for 25 iterations. Figure 6.9 shows the integration graph of the SMIB system. Table 6.2 and Table 6.3 present the parameters of PSS for various X_e values.

Table 6.4 shows the steadiness of a system with high negative Eigen values. It is understood from the analysis that AGPSO1 is the best of all the PSO variants. It is also designed to swap the actual response of the network due to interference. The adjusted PSS parameters for network changes were extracted from the previous section. With the network parameter $X_e = 0.2$, the below cases are observed.

Case (i): V_{ref} variations
Case (ii): T_{mech} variations for increment in the load
Case (iii): T_{mech} variations for decrement in the load

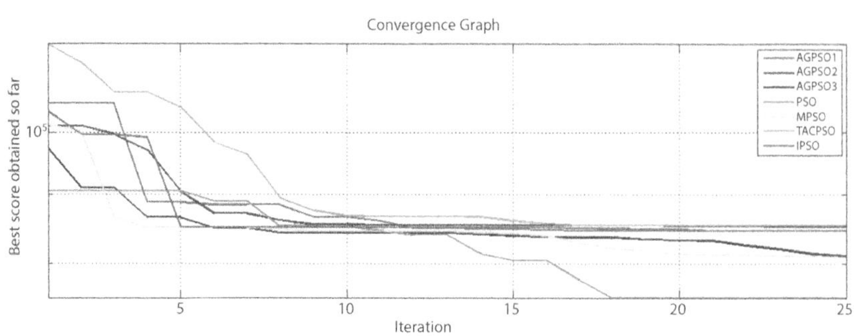

Figure 6.9 Convergence – SMIB system.

Table 6.2 Parameters of PSS - optimization algorithms.

Type/ X_E	AG-PSO1				AG-PSO2				AG-PSO3				SVM				
	K_S	T_W	T_1	T_2	K_S	T_W	T_1	T_2	K_S	T_W	T_1	T_2	C & γ	K_S	T_W	T_1	T_2
0.2	5	15	0.707899	0.228489	5	15	0.961589	0.288118	9.13421	15	0.239761	0.106107	0.0001	11	15	0.998	0.35
0.4752	22.4412	15	0.334591	0.284708	22.957	14.9914	0.247676	0.209731	21.4371	15	0.624424	0.520233	0.001	10.5	15	1	0.8
0.65	29.8752	15	0.665415	0.644167	31.6392	15	0.976853	1	30.8318	15	1	1	0.01	10.45	15	1	1
0.8	36.0371	15	0.941299	1	36.1199	15	0.9398	1	33.3016	15	1	1	0.1	10.45	15	1	1
1.08	50	15	1	0.1	50	15	1	0.1	50	15	1	0.1	1	10.45	15	1	1
1.2	50	15	1	0.1	50	15	1	0.1	50	15	1	0.1	10	10.45	15	1	1
1.4	50	15	1	0.1	50	15	1	0.1	50	15	1	0.1	100	10.45	15	1	1

Table 6.3 Parameters of PSS - optimization algorithms.

Algo/ X_e	TACPSO				IPSO				MPSO			
	K_S	T_W	T_1	T_2	K_S	T_W	T_1	T_2	K_S	T_W	T_1	T_2
0.20	5.0	14.95	0.700	0.227	5.0	15.0	1.0	0.296	9.602	15.0	0.256	0.122
0.470	21.30	14.99	0.340	0.284	22.54	14.99	0.182	0.150	22.76	15.0	0.249	0.211
0.650	29.26	15.0	0.659	0.632	30.82	15.0	1.0	1.0	31.67	15.0	0.976	1.0
0.80	36.04	15.0	0.940	1.0	33.30	15.0	1.0	1.0	36.07	15.0	0.940	1.0
1.08	50.0	15.0	1.0	0.10	50.0	15.0	1.0	0.10	50.0	15.0	1.0	0.10
1.2	50.0	15.0	1.0	0.10	50.0	15.0	1.0	0.10	50.0	15.0	1.0	0.10
1.4	50.0	15.0	1.0	0.10	50.0	15.0	1.0	0.10	50.0	15.0	1.0	0.10

Case (i): In first case, the preliminary condition is to fix the reference voltage as 1 p.u. After some time, let's assume at t = 20 seconds, the reference voltage varies from 1 p.u. to 1.267 per unit using a knob which is present on the Control Desk. The mechanical outcome of these variables is depicted in Figure 6.10a.

Case (ii): In the second case, the torque of the load equipment varies by multiplying the load variation in the PS. The T_{mech} is initially set at 1 per unit and then switched to 0.7790per unit during t=20 seconds. The machine responses are illustrated in Figure 6.10b. From this figure, it is clearly observable that the power of the terminal is sustained regardless of the variation of the input torque as the machine is attached to the endless bus. As the Tm is decreased, the Te also decreases. Vibrations in rotor speed outcomes die faster due to the existence of a PS.

Case (iii): In the third case, the torque of the equipment has been raised to 1.242 per unit during t=20 seconds. The variations of the machine parameters are illustrated in Figure 6.10c.

Table 6.4 Eigen values for test system.

AG-PSO1	AG-PSO2	AG-PSO3	PSO	IPSO	TAC-PSO	MPSO
X_E=0.4752						
-35.4493 -4.3474 +11.2007i -4.3474 -11.2007i -0.0673 -4.8739 + 0.7107i -4.8739 - 0.7107i	-35.5122 -4.5569 +11.2918i -4.5569 -11.2918i -0.0673 -5.2607 + 2.0794i -5.2607 - 2.0794i	-35.3935 -4.1442 +11.1215i -4.1442 -11.1215i -6.5331 -0.0672 -2.0868	-35.5329 -4.1901 +11.7589i -4.1901 -11.7589i -6.4742 -0.0249 -0.7572	-35.3898 -4.2515 +10.9596i -4.2515 -10.9596i -0.0672 -4.9994 + 0.6328i -4.9994 - 0.6328i	-35.5535 -4.7957 +11.1056i -4.7957 -11.1056i -0.0673 -5.9392 + 3.1399i -5.9392 - 3.1399i	-35.4957 -4.5224 +11.2334i -4.5224 -11.2334i -0.0673 -5.2846 + 2.0465i -5.2846 - 2.0465i
X_E=0.65						
-32.5078 -4.7535 +10.0725i -4.7535 -10.0725i -8.3017 -0.0674 -1.5712	-32.5052 -4.7025 +10.1044i -4.7025 -10.1044i -8.4304 -0.0674 -0.9948	-32.5000 -4.7079 +10.0808i -4.7079 -10.0808i -8.4194 -0.0674 -1.0000	-32.5971 -4.8315 +10.3465i -4.8315 -10.3465i -8.0770 -0.0166 -0.2153	-32.4795 -4.7114 + 9.9964i -4.7114 - 9.9964i -8.4070 -0.0674 -1.6081	-32.4998 -4.7075 +10.0803i -4.7075 -10.0803i -8.4203 -0.0674 -1.0000	-32.5058 -4.7030 +10.1062i -4.7030 -10.1062i -8.4288 -0.0674 -0.9946

(Continued)

Table 6.4 Eigen values for test system. (*Continued*)

AG-PSO1	AG-PSO2	AG-PSO3	PSO	IPSO	TAC-PSO	MPSO
X_E=1.08						
-8.8102 +43.0373i -8.8102 -43.0373i -42.1298 1.6055 -0.0656 -2.1315	-8.8102 +43.0373i -8.8102 -43.0373i -42.1298 1.6055 -0.0656 -2.1315	-8.8102 +43.0373i -8.8102 -43.0373i -42.1298 1.6055 -0.0656 -2.1315	1.0e+03 * -0.4944 +4.3320i -0.4944 - 4.3320i -0.0500 0.0002 -0.0002 -0.0000	-8.8102 +43.0373i -8.8102 -43.0373i -42.1298 1.6055 -0.0656 -2.1315	-8.8102 +43.0373i -8.8102 -43.0373i -42.1298 1.6055 -0.0656 -2.1315	-8.8102 +43.0373i -8.8102 -43.0373i -42.1298 1.6055 -0.0656 -2.1315

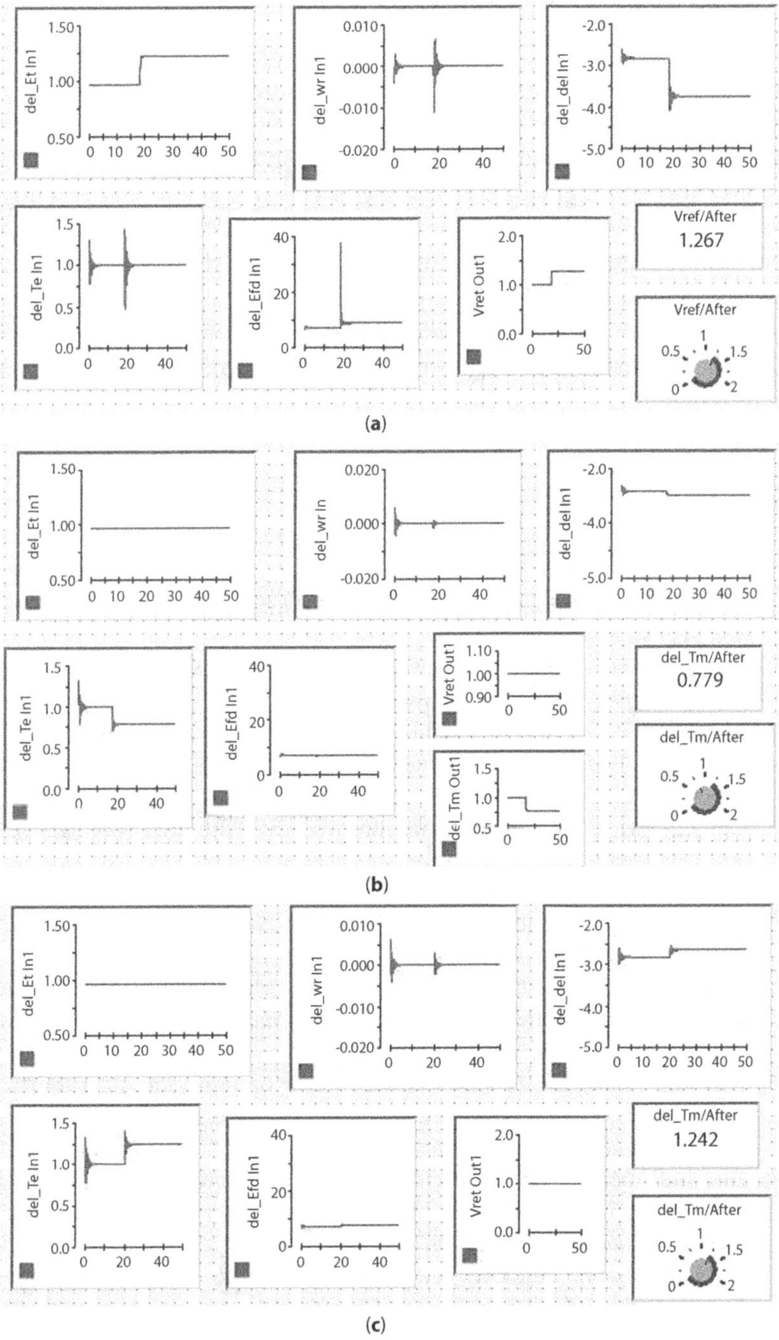

Figure 6.10 (a) Responses of machine variables - change in V_{ref} for system with X_e=0.2. (b) Responses of machine parameters - change in T_{mech} with X_e=0.2. (c) Responses of machine variables -change in T_{mech} for the system with X_e=0.2.

6.6 Conclusion

The results of the optimization algorithm study using PSO and SVM responsible for identifying the appropriate control for the various operating conditions were studied and analysed. An excellent set of points for various models designed for modelling is available and a relative study is conducted on the production outcome of seven contrasting performance algorithms set out for the two Single Machine and Multi Machine System power systems. Excellent scores are tested and verified in a real-time power system using the dSPACE simulator. This form of real-time analysis helps to expand the installed power system in the future without adversely affecting its reliability and protection.

References

1. Demello, FP and Concordia, C (1969). Concepts of synchronous machine stability as affected by excitation control. *IEEE Trans. on Power Apparatus and Systems*,88(4),316-329.
2. Pai, MA, Sen Gupta, DP &Padiyar, KR. Small Signal Analysis of Power Systems. *Alpha Science International Ltd.*2004.
3. Kundur, P. 'Power System Stability and Control. McGraw-Hill, Inc., 1994.
4. Anderson, P.M. and A.A. Fouad. Power System Control and Stability. *GalGotia Publications*,1981.
5. Cui, Z, Zeng, J, Yin, Y. (2008). An improved PSO with time-varying accelerator coefficients. *Eighth international conference on intelligent systems design and applications*, 638-643.
6. Eberhart R.C, Kennedy, J (1995). Particle Swarm Optimization. *IEEE International Conference on Neural Network*, 4,1942-1948
7. Ziyu, T.; Dingxue, Z (2009). A modified particle swarm optimization with an adaptive acceleration coefficients. *Asia-Pacific Conference on Information Processing, Shenzhen*,330- 332.
8. Bao, G.Q, Mao, K.F (2009). Particle swarm optimization algorithm with asymmetric time varying acceleration coefficients. *IEEE International Conference on Robotics and Biomimetic*,2134-2139.
9. Alrashidi, MR & El-Hawary, ME (2006). A Survey of Particle Swarm Optimization Application in Electric Power Systems. *Journal of Electric Power Components and systems*, 34,1349-1357.
10. SayedaliMirjalili, Andrew Lewis, Ali Safa Sadiq (2014). Autonomous Particles Groups for Particle Swarm Optimization. *Arabian Journal of Science and Engineering*,39(6),4683–4697.

11. Chee-Mun Ong. Dynamic Simulation of Electric Machinery – using MATLAB Simulink. Prentice Hall, 1997.
12. dSPACE Catalog. DS1104 R & D controller board hardware installation and configuration for DS1104.Paderborn, Germany,2009.
13. K.Selvi and R.Ramya (2010). Real Time Simulation of Synchronous Machine using dSPACE'. *Journal for Institution of Engineers*, 91,29-36.
14. R.Ramya and K.Selvi (2014). Real Time simulation of Single Machine Infinite Bus system using dSPACE controller board. *Lecture Notes in Electrical Engineering series of Springer*. 326, 783-792.
15. R.Ramya, K.Selvi and K.Murali (2016). Real time simulation and performance analysis of Multi machine power system using dSPACE simulator. *Simulation-Transaction of the society for modelling and simulation international*, 92(1), 63-75.
16. Suganyadevi, M. V., and C. K. Babulal (2014). Support vector regression model for the prediction of loadability margin of a power system. *Applied Soft Computing* 24 ,304-315.
17. Suganyadevi, M.V., Babulal, C.K. and Kalyani, S (2016). Assessment of voltage stability margin by comparing various support vector regression models. *Soft computing*, 20(2), 807-818.

7

Modelling and Control of PMSM Drives

Souvik Ganguli[1*], Yashonidhi Srivastava[1], Gagandeep Kaur[1] and Prasanta Sarkar[2]

[1]*Electrical and Instrumentation Engineering Department, Thapar Institute of Engineering and Technology, Patiala, Punjab, India*
[2]*Electrical Engineering Department, National Institute of Technical Teachers' Training & Research, Kolkata, West Bengal, India*

Abstract
With regards to the complex delta domain, a hybrid metaheuristic algorithm is used to decrease the model orders and design the controller of the permanent magnet synchronous motor, commonly known as PMSM. Two popular algorithms viz. the firefly technique abbreviated as FA and the pattern search (PS) technique are combined to develop a new topology termed the hybrid firefly algorithm, designated in this chapter as the FAPS method. The FA technique performs a global search while the PS method is utilized for local searches in the hybrid algorithm. In the beginning, a PMSM drive consisting of both the speed and current controllers was reduced using the FAPS technique to produce a second-order model. In a cascade configuration with a proportional-integral (PI) controller, the lower-order system is further compared with the desired system to identify roughly the undiscovered controller parameters. The values of the controller gains obtained using the delta transform almost imitate that observed in the continuous-time mode. Thus, a unified controller design framework is constructed for the drive. The proposed algorithm is therefore used effectively for both order diminition and the measurement of controller parameters for PMSM drives. Besides, for many domestic and industrial uses, a case study to regulate some special motor drives may also be studied in the future.

***Keywords*:** Modelling and control, firefly technique, electric drive, permanent magnet synchronous motor, unified delta domain, pattern search algorithm

**Corresponding author*: souvik.ganguli@thapar.edu: ORCID: https://orcid.org/0000-0003-4192-398X

Vipin Chandra Pal, Suman Lata Tripathi, and Souvik Ganguli (eds.) Industrial Control Systems, (137–150) © 2024 Scrivener Publishing LLC

7.1 Introduction

The commonly used control applications in the automation and aviation industries quite often generate mathematical models of a higher order. Conceptualizing the development of higher-order controllers for these systems are rather difficult. Therefore, these systems are first of all reduced to help minimize the computational burden. Moreover, the problems associated with implementing the higher-order controllers should also be decreased considerably in the process. Although literature lists several intelligent control techniques for PMSM [1, 2], the PID control system prevails to entice industrial applications owing to its elementary form and strength in a variety of operations. Some significant literature research work that will generate interest among the researchers in the speed control of PMSM drives are enumerated below.

A model referring to an adaptive control system was presented utilizing the fuzzy logic adaptation mechanism. The adaptation mechanism was adopted to generate additional signals for the input of the system. The model introduced pertaining to the angular system concerned with a (PMSM) drive was implemented onto the linearized model which was magnet synchronous and vector controlled. In the case of massive variations applied to the parameters, the simulation provided promising adaptations. The system was also compatible with the microcomputer implementation [3].

A flux weakening method was introduced to be implemented at the interior portion of the permanent synchronous motor. The method proposed worked on the basis of the output of the PI regulator. The output voltage regulation loop was responsible for weakening the flux as well as regulating its levels to inhibit the saturation of the current regulator. The feasibility and the ability of the system were introduced to ensure current regulation was confirmed and backed by fast transitions [4].

Aimed for practical implementations, a self-constructing map constituting fuzzy logic and the neural net was presented. The structure happens to be working on the segregation of input spacing in addition to other factors to successfully monitor the gradient descent method utilizing the law of delta adaption. The efficiency of the proposed method was backed by multiple experimental outcomes along with the accomplishment of a PMSM drive. The outputs generated from the simulations of the nonlinear disturbances paired with time-varying complexity were utilized to provide the traits of the controller under several operating conditions [5].

A study on a sensorless control method was carried out to assess the design, modelling, and control for a permanent magnet synchronous model. The model was based on the Kalman filter. The functionality was tested for several dynamic as well as static loads. It was facilitated by starting the motor at distinct rotor positions. In the case of parameter variations in PMSM, it turned out to be insensitive to the proposed extended Kalman filtering estimation method. The performance and operations were verified through experiments [6].

A mathematical model for a sensorless vector control indulged in a (PMSM) machine was presented. A motor drive encompassing three-phase PWM inverters as well as PMSM control systems were introduced emphasizing the frequency response based designs pertaining to the associated loops. Some of those presented were rotor speed and position estimator, speed controllers, and back-EMF state observer. The performance of the presented model was verified and backed by experimental results conducted with a PMSM drive [7].

The development of a sensorless, PMSM drive was discussed. An observer-based approach, namely the Luenberger observer was applied to determine the position of the rotor flux and further assess the rotor speed. The simple structure and the absence of mechanical parameters made the motor additionally efficient. With the intention of the determination of the rotor speed, a feedback method was adopted. Several simulations and experiments along with operations at low speeds verified the feasibility and efficiency of the system [8].

A speed estimation methodology was presented for the PMSM drives. The formulation of an adaptive system based on a reference model utilizing the reactive powers under steady-state conditions was discussed. The presence of numerous distinctive features also discoursed along with the fact that the system was independent of stator resistance. Popov's Hyperstability criterion was adopted to gain stability in the system. The test system was validated at several distinct speeds with the inclusion of even zero speed. The results and outcomes of the simulation were presented that ensure the applicability of the methodology under discussion [9].

Methodologies to minimize the occurrence of both the flux as well as the torque ripples in direct torque controlled (DTC) PMSM drives have been showcased. The execution of duty determination methods was discussed and an effective, robust yet simple method was presented. The minimization of commutation frequency and the arrangement of vectors improved the results of the system. The proposed method significantly reduced the flux ripple and torque, which was backed by experimental and simulation results [10].

A fuzzy logic-based approach was introduced to reduce the chaos feature of the electric drive utilizing the backstepping technique. The backstepping method helped in the implementation of construct controllers while the fuzzy logic systems were used to determine nonlinearities that happened to remain unknown. The structure presented has been kept simple yet resulted in an effective outcome. The model was successfully able to reduce the chaos in the system and track the reference signal when faced with uncertain parameters [11].

The applications of fractional order systems with reference to the modelling of the PMSM velocity servo techniques were presented. The transformation of the traditional model to the one having a fractional order has been described. An integer approximation technique was employed to determine the factors of the system expressed in fractional-order. The structure in state-space was then presented utilizing an output-error identification algorithm. By utilizing a scheme based on the identified fractional-order model, two proportional controllers were designed. Several experiments were conducted utilizing the controllers. The results highlighted the benefits of the proposed system as well as proved its deductions to be accurate and trusty [12].

An improved technique of MPCC for the PMSM drives has been introduced. The technique first estimates the electromotive force depending on previous values and then applied the estimated EMF in the stator current that was predicted. A null vector was also obtained to reach a steady-state performance during the application of MPCC. Two techniques were proposed requiring six and one predictions, respectively, to attain optimum results [13].

The methodology of adaptive backstepping optimal control (ABOC) was introduced to attain good performance from PMSM drives undergoing linear motion with a ball screw. The design of the ABOC system, as well as its implementation, had been discussed. The technique was described as an amalgamation of a backstepping controller with a robust and optimal controller. Experimental results have been included that confirm the technique's performance in terms of tracking [14].

The model predictive control approach used for regulating the current had been popularly used for the purpose of controlling the PMSM drives. An advanced and improved version was presented in this paper. The advancements included the implementation of an active VV in addition to a zero VV to the duty cycle control (DCC). The method proposed was based on the calculations of the slopes due to several different VVs. Comparisons with MPCC, C-MPCC, and DCC have been included that verify the effectiveness and accuracy of the system [15].

It can be rightly said that almost all of the discrete systems are derived from continuous system sampling. The modelling and analysis as well as control of those systems are executed by a time-domain approach or a complex domain z-transform analysis. These systems do not, therefore, take into consideration an exceptionally high sampling rate and are not stabilized at close range due to pole crowding near (1,0). This can be achieved if the Middleton and Goodwin-created delta operator is used to formulate the discrete-time systems [16]. Simulating these systems in the delta domain with a high sampling frequency helps to construct a replica of the continuous-time systems and simultaneously increases computational control of the discrete-time systems developed keeping the shift operator as the foundation. The delta transform-based systems are now commonly used for research and design of control systems [17–20].

The Truxal approach is a long-standing synthesis based approach and has been used widely for control applications [21]. This works on the principle of precise model matching where the reference model is first configured to satisfy certain time and frequency domain constraints. The controller gains are thus calculated when the response of the complete controlled plant model in a closed-loop form that coincides with the reference model. The most critical demerit of the exact matching model (EMM) [22] is that the design does not ensure that the controller is physically implemented. Approximate model matching abbreviated as AMM [23] can be an alternate approach to solve the issue and could be extended to the design control system in the united domain of work. A hybrid firefly method coined as FAPS was put forward by Ganguli *et al.* [24] to lower the higher-order models in the continuous-time framework. This algorithm has been employed in this chapter using the AMM framework to diminish the third-order drive system and further tuning the controller gains in the complex delta domain.

The remaining chapter is designed as follows. Section 7.2 provides an outline of the suggested work. Section 7.3 addresses the simulation outcomes and Section 7.4 explores the key inferences deliberating some issues which will be taken up in future.

7.2 A Proposed Technique for Modelling and Control

This chapter examines the modelling aspects and controller design problem of a PMSM drive system. The assigned job is therefore segregated into two sections. The first section covers the order diminution of the drive model considered from the literature. The PMSM drive presents a higher-order system when modelled. However, to build the controller, there is a requirement of a low order model for implementation issues of the controller. A hybrid

FAPS [24] technique developed using the firefly algorithm combined with pattern search (PS) is used to reduce the parent PMSM drive model.

The authors of this paper developed this hybrid algorithm, which integrated FA with Pattern Search (PS) for reduced-order modelling of continuous-time systems [33]. By applying the firefly technique as a global optimizer to conduct exploration, while pattern search best suited for a local search to provide an intensification function, the perfect inclination towards both exploitation as well as exploration was achieved. The algorithm utilizes the convenience of the firefly technique and pattern search methods truly and avoids their drawbacks. In the FAPS technique, a collective set of random variables helped to boot the search technique applying the firefly method. Thus, the firefly algorithm executed the search for one-half of the entire iterations. The solving done by the firefly method was used as an initial point for the local search approach carried out by the pattern search algorithm. The hunt to find the feasible solution was then dragged in the direction of the pattern search technique to accelerate the global convergence towards an acceptable outcome. Therefore, the hybrid approach tried for an optimal solution faster and with better accuracy.

The coefficients of the numerator and the denominator were both considered as unknown and were calculated using an optimized FAPS algorithm applying constraints to maintain DC gain, minimum phase, and stability. A random sequence, popularly coined as PRBS in the literature, is used to obtain the responses of both the original and reduced-order models containing unknown numerator and denominator terms. The difference between them is regarded as an error. To get the numerator and denominator polynomial, the sum of square error (SSE) is minimized. The second part of the work involves designing a controller where three-term controller coefficients are calculated with the FAPS method. The controller's synthesis is carried out by the concept of approximate model matching (AMM). Numerous recent heuristic techniques are used to justify the effectiveness of the proposed methodology, such as chicken swarm optimization (CSO), atomic search optimization (ASO), marine predators algorithm (MPA), slime mold algorithm (MFO), and a hybrid algorithm constituting a grey wolf optimizer (GWO) and cuckoo search (CS), and nomenclature as GWOCS [26–31] in addition to the parent firefly algorithm (FA) [25].

Also, in comparison with the original system, several time-domain parameters of the reduced models are evaluated with respect to the original system. A comparison of some of the recently evolved metaheuristic methods is evaluated with five benchmark error output indices. Thus, the controller parameters in the approximate model matching (AMM) framework are calculated using FAPS. The response from the plant control cascade and the selected reference model is compared to the controller synthesis problem to minimize the integral square error (ISE) to evaluate controller gains. Besides, there is also a minimum fitness value.

7.3 Results and Discussions

The transfer function of the PMSM drive is given by

$$G_{cont}(s)=\frac{7.94s+13.23}{9.6\times10^{-7}s^3+3.99\times10^{-3}s^2+6.97s+34.63} \tag{7.1}$$

Our ultimate purpose is to create a suitable PMSM drive controller. As the input-output relationship generated from the information provided in [31, 32] is of order three, the initial stride is to convert it into a second-order system suitable for standard PID controllers. Rather, a proportional-integral control action is normally picked in place of a benchmark controller, usually PID, in the case of drive operation because it can be used during the drive process to prevent major disturbances and noise.

For the order reduction, the PMSM drive model represented in Equation (7.1) is converted to the discrete form using a suitable delta transform. A sampling time of 0.001 seconds is chosen for this purpose. Thus, the system in the delta operator framework is represented as

$$G_{delta}(\gamma)=\frac{100\gamma^2+1200300\gamma+1995900}{\gamma^3+2040\gamma^2+1061200\gamma+5.224200} \tag{7.2}$$

In the delta domain with acceptable population sizes and maximum iterations of 20 and 100, the reduced models are obtained. Standard parameters of all algorithms (proposed, parent, and the latest heuristics) are taken into account as described in the literature. The reduced systems in the mixed domain are presented in Table 7.1. In the same table, the minimum fitness values are also presented. The bolded outcome in the table shows the best result.

The recommended FAPS technique supersedes the constituent algorithms. Further, improvement is also observed as compared to some benchmark heuristic approaches when referred to some popular and recent literature records. Moreover, the suggested approach generates a minimum

Table 7.1 Model reduction and minimum cost function of drive model in fused domain of work.

Applied methods	Reduced systems using delta transform	Minimum cost function
FAPS (Proposed)	$\frac{997.8\gamma+1352.0508}{\gamma^2+851.9\gamma+3539.4}$	**2.9695e-04**
FA [25]	$\frac{989.4\gamma+2435.2118}{\gamma^2+843.8\gamma+6374.9}$	3.5926e-04
CSO [26]	$\frac{991.01\gamma+2187.142}{\gamma^2+844.87\gamma+5727.48}$	3.2801e-04
ASO [27]	$\frac{992.34\gamma+2156.3235}{\gamma^2+846.8798\gamma+5644.7781}$	4.9872e-04
MPA [28]	$\frac{997.5112\gamma+1468.8125}{\gamma^2+835.6214\gamma+3844.9899}$	3.3887e-04
SMA [29]	$\frac{954.25\gamma+1994.9879}{\gamma^2+836.4113\gamma+5224.1982}$	5.2283e-04
GWOCS [30]	$\frac{991.62\gamma+3786.2}{\gamma^2+788.512\gamma+9911.4892}$	3.0722e-04

fitness value, as shown in Table 7.1. Only the hybrid method, namely the GWOCS technique gives an equally competitive result. Moreover, the time domain parameters are calculated for the reduced models, the results of which are showcased in Table 7.2. The rise time (t_r) and the settling time (t_s) are indicated in seconds (s), while the peak overshoot (M_p) is denoted with the help of a percentage. Only the time domain parameters are taken into account. Similarly, the frequency domain parameters can also equally be explored in subsequent times. The bold results shown in the table are nearest to the original parameters of the model.

Compared to the other algorithms in this chapter, the hybrid technique again works better. However, the closest settling time is provided by the MPA method. The outcomes of CSO and ASO are quite competitive.

Besides this, for the reduced-order models illustrated in Table 7.2, the error metrics commonly used in system theory and control are evaluated. The results are shown in Table 7.3. The selected algorithms are used to juxtapose the results. The best result is the lowest possible value with respect to any benchmark metric listed in the table, which is indicated by using striking letters to distinguish them from the other outcomes.

The results in Table 7.3 suggest that many of the recent heuristic approaches have been outperformed by the proposed method. The next step in the unified delta domain is to determine the PI controller gains. As discussed earlier, regarding controller synthesis, the AMM technique is used. The controller parameters can be heuristically calculated using the algorithm proposed, whose results are recorded in Table 7.4. The best cost function is also found by the heuristic scheme besides the control parameters. The minimum fitness value is again indicated utilizing bold letters.

In comparison to some existing heuristic strategies, the proposed hybrid algorithm achieves the lowest optimal value although the latest techniques like ASO and MPA show quite a close match. The proposed method could therefore successfully generate PMSM models of a lower order and use the approximate model matching technique to create a PI controller that has the lowest fitness function compared to quite a few existing algorithms.

Table 7.2 Time-domain specifications of lower order electric drive model applying the undivided domain method.

PMSM model	Methods	t_r (in secs)	t_s (in secs)	M_p (%)
Parent model		0.0027	0.7803	204.4378
Second-order Models	Proposed Method	**0.0003**	0.9449	**202.3298**
	FA [25]	0.0004	0.5205	200.1620
	CSO [26]	3.7088e-04	0.5799	200.9022
	ASO [27]	3.7061e-04	0.5897	200.5786
	MPA [28]	3.6745e-04	**0.8534**	207.6500
	SMA [29]	3.8711e-04	0.6292	193.0724
	GWOCS [30]	3.6575e-04	0.3133	217.7588

Table 7.3 Performance metrics of lower-order electric drive system in combined domain.

Methods	Integral of absolute error	Integral of time-weighted absolute error	Integral of square error	Integral of time-weighted square error	Norm-based error
Proposed	**0.0055**	**3.4966e-04**	**2.9695e-04**	**2.0060e-05**	**0.0754**
CSO [26]	0.0405	0.0025	0.0175	0.0011	0.6693
ASO [27]	0.0404	0.0025	0.0174	0.0011	0.6679
MPA [28]	0.0399	0.0025	0.0173	0.0011	0.6841
SMA [29]	0.0431	0.0027	0.0191	0.0012	0.6854
GWOCS [30]	0.0398	0.0025	0.0177	0.0011	0.7103

Table 7.4 Values of controller gains and cost function for reduced-order electric drive model in consolidated domain.

Methods	K_P	K_I	Fitness function value
FAPS	1.8679	1.2146e-05	**7.3715e-05**
CSO [26]	0.0029	4.0769	7.9023e-05
ASO [27]	0.0028	4.0689	7.4273e-05
MPA [28]	0.0037	3.7721	7.3987e-05
SMA [29]	0.0039	4.1267	7.7577e-05
GWOCS [30]	0.0010	4.3866	7.6758e-05

7.4 Conclusions

This chapter discusses the subject of order diminution and development of the controller for an electric drive in the discrete-delta domain. A permanent magnet AC motor drive is considered for this work. Specifically, a synchronous motor is taken up for investigation. The reduced system for the electric drive is being formulated with a global optimization technique applying suitable constraints. For this reason, a hybrid topology combining the pattern search with the firefly algorithm is successfully used. Also, the method uses the AMM method to obtain the controller gain values. A PI controller is thus implemented. The advantage is that this approach uses only output feedback. The control system is designed to follow the desired model closely. In this chapter, an important application to model and speed control the PMSM drive in the unified delta domain has been efficiently discussed. Similar modelling and control applications can also be done for some new and upcoming drives such as the switched reluctance motor and BLDC drive systems.

References

1. Ibrahim, Z., & Levi, E. (2002). A comparative analysis of fuzzy logic and PI speed control in high-performance AC drives using experimental approach. *IEEE Transactions on Industry Applications*, *38*(5), 1210-1218.
2. Yu, J., Shi, P., Dong, W., Chen, B., & Lin, C. (2015). Neural network-based adaptive dynamic surface control for permanent magnet synchronous motors. *IEEE Trans. Neural Netw. Learning Syst.*, *26*(3), 640-645.

3. Kovacic, Z., Bogdan, S., &Crnosija, P. (1993, November). Fuzzy rule-based model reference adaptive control of permanent magnet synchronous motor drive. In *Proceedings of IECON'93-19th Annual Conference of IEEE Industrial Electronics* (pp. 207-212). IEEE.
4. Kim, J. M., & Sul, S. K. (1997). Speed control of interior permanent magnet synchronous motor drive for the flux weakening operation. *IEEE Transactions on Industry Applications, 33*(1), 43-48.
5. Lin, F. J., Lin, C. H., & Shen, P. H. (2001). Self-constructing fuzzy neural network speed controller for permanent-magnet synchronous motor drive. *IEEE transactions on fuzzy systems, 9*(5), 751-759.
6. Qiu, A., Wu, B., &Kojori, H. (2004, May). Sensorless control of permanent magnet synchronous motor using extended Kalman filter. In *Canadian Conference on Electrical and Computer Engineering 2004 (IEEE Cat. No. 04CH37513)* (Vol. 3, pp. 1557-1562). IEEE.
7. Burgos, R. P., Kshirsagar, P., Lidozzi, A., Wang, F., &Boroyevich, D. (2006, July). Mathematical model and control design for sensorless vector control of permanent magnet synchronous machines. In *2006 IEEE Workshops on Computers in Power Electronics* (pp. 76-82). IEEE.
8. Chan, T. F., Wang, W., Borsje, P., Wong, Y. K., & Ho, S. L. (2008). Sensorless permanent-magnet synchronous motor drive using a reduced-order rotor flux observer. *IET Electric Power Applications, 2*(2), 88-98.
9. Maiti, S., Chakraborty, C., & Sengupta, S. (2009). Simulation studies on model reference adaptive controller based speed estimation technique for the vector controlled permanent magnet synchronous motor drive. *Simulation Modelling Practice and Theory, 17*(4), 585-596.
10. Zhang, Y., & Zhu, J. (2010). Direct torque control of permanent magnet synchronous motor with reduced torque ripple and commutation frequency. *IEEE Transactions on Power Electronics, 26*(1), 235-248.
11. Yu, J., Chen, B., Yu, H., & Gao, J. (2011). Adaptive fuzzy tracking control for the chaotic permanent magnet synchronous motor drive system via backstepping. *Nonlinear Analysis: Real World Applications, 12*(1), 671-681.
12. Yu, W., Luo, Y., & Pi, Y. (2013). Fractional order modeling and control for permanent magnet synchronous motor velocity servo system. *Mechatronics, 23*(7), 813-820.
13. Zhang, Y., Xu, D., Liu, J., Gao, S., & Xu, W. (2017). Performance improvement of model-predictive current control of permanent magnet synchronous motor drives. *IEEE Transactions on Industry Applications, 53*(4), 3683-3695.
14. El-Sousy, F. F., Amin, M., & Mohammed, O. A. (2019). Robust Optimal Control of High-Speed Permanent-Magnet Synchronous Motor Drives via Self-Constructing Fuzzy Wavelet Neural Network. In *2019 IEEE Industry Applications Society Annual Meeting* (pp. 1-8). IEEE.
15. Petkar, S. G., Eshwar, K., & Thippiripati, V. K. (2020). A Modified Model Predictive Current Control of Permanent Magnet Synchronous Motor Drive. *IEEE Transactions on Industrial Electronics.*

16. Middleton, R. H., & Goodwin, G. C. (1990). *Digital Control and Estimation: A Unified Approach (Prentice Hall Information and System Sciences Series)*. Englewood Cliffs, NJ: Prentice Hall.
17. Wang, J., Gao, Y., Qiu, J., & Ahn, C. K. (2016). Sliding mode control for non-linear systems by Takagi–Sugeno fuzzy model and delta operator approaches. *IET Control Theory & Applications, 11*(8), 1205-1213.
18. Huang, C., Naghdy, F., & Du, H. (2017). Delta operator-based fault estimation and fault-tolerant model predictive control for steer-by-wire systems. *IEEE Transactions on Control Systems Technology, 26*(5), 1810-1817.
19. Kumari, K., Bandyopadhyay, B., Kim, K. S., & Shim, H. (2019). Output feedback based event-triggered sliding mode control for delta operator systems. *Automatica, 103*, 1-10.
20. Ramírez-Neria, M., Luviano-Juárez, A., Lozada-Castillo, N., Ochoa-Ortega, G., & Madonski, R. (2020). Discrete-Time Active Disturbance Rejection Control: A Delta Operator Approach. In *Advanced, Contemporary Control* (pp. 1383-1395). Springer, Cham.
21. Truxal, G. J. (1955). Automatic Feedback Control System Design. McGraw-Hill, New York.
22. Wolovich, W. A. (1972). The use of state feedback for exact model matching. *SIAM Journal on Control, 10*(3), 512-523.
23. Wang, S.S., & Chen, B.S. (1987). Model matching: optimal approximation approach. *International journal of control, 45*(6), 1899-1907.
24. Ganguli, S., Kaur, G., & Sarkar, P. (2016). Model order reduction of continuous time system using hybrid metaheuristic algorithm. In IEEE *7th India International Conference on Power Electronics (IICPE)*, (pp. 1-5).
25. Yang, X. S. (2010). Firefly algorithm, stochastic test functions and design optimisation. *International journal of bio-inspired computation, 2*(2), 78-84.
26. Meng, X., Liu, Y., Gao, X., & Zhang, H. (2014). A new bio-inspired algorithm: chicken swarm optimization. In *International conference in swarm intelligence* (pp. 86-94). Springer, Cham.
27. Zhao, W., Wang, L., & Zhang, Z. (2019). A novel atom search optimization for dispersion coefficient estimation in groundwater. *Future Generation Computer Systems, 91*, 601-610.
28. Faramarzi, A., Heidarinejad, M., Mirjalili, S., & Gandomi, A. H. (2020). Marine Predators Algorithm: A nature-inspired metaheuristic. Expert Systems with Applications, 152, 113377.
29. Li, S., Chen, H., Wang, M., Heidari, A. A., & Mirjalili, S. (2020). Slime mould algorithm: A new method for stochastic optimization. Future Generation Computer Systems, 111, 300-323.
30. Long, W., Cai, S., Jiao, J., Xu, M., & Wu, T. (2020). A new hybrid algorithm based on grey wolf optimizer and cuckoo search for parameter extraction of solar photovoltaic models. Energy Conversion and Management, 203, 112243.

31. Ayyar, K. (2013). Investigations on model order reduction and design of speed controller for electric drive systems.
32. Krishnan, R. (2001). *Electric motor drives: modeling, analysis, and control* (Vol. 626). New Jersey: Prentice Hall.
33. Ganguli, S., Kaur, G., & Sarkar, P. (2016). Model order reduction of continuous time system using hybrid metaheuristic algorithm. In 2016 7th India International Conference on Power Electronics (IICPE), pp. 1-5.

8

VI System for Power Management of DC Microgrid

Roop Pahuja* **and Ranveer Singh**

Dept. of Instrumentation & Control Engineering, Dr. B R Ambedkar National Institute of Technology, Jalandhar (Punjab), India

Abstract

To overcome today's energy crisis problem, a more realistic and consumer-friendly approach to the utility grid system is a microgrid. Motivated to re-design some of the essential subsystems of a DC microgrid for efficient power management in a more informative and user-friendly manner, this chapter presents a virtual instrumentation metering infrastructure and a modern ICS (Industrial Control System) for the same. The proposed microgrid system backed by different power resources such as renewable solar energy, a battery backup, and a rectified grid uses a distributed embedded data acquisition-communication hardware and software as metering components for its power management. A finite state machine supervisory controller is implemented to optimize the power utilization of microgrid resources and maintain a balance between demand and supply of electrical energy. The system has a customized microgrid remote user interface that displays all the vital parameters such as the operating status of a microgrid, power utilization of different energy resources, energy consumption, load selection, and a temperature-based health alarm for the solar panel. The system prototype is successfully developed and tested under many real-time energy-load conditions with promising results. Such a system offers a cost-effective and scalable solution for effective power management and monitoring of a small DC microgrids in homes, shops, small buildings, or remote areas.

Keywords**:** DC microgrid, power management, finite state machine supervisory controller, software metering panel for microgrid, remote user interface, virtual instrumentation, modern industrial control system

**Corresponding author*: pahujar@nitj.ac.in

Vipin Chandra Pal, Suman Lata Tripathi, and Souvik Ganguli (eds.) Industrial Control Systems, (151–180) © 2024 Scrivener Publishing LLC

8.1 Introduction

Industrial Control System

The proposed system for a DC microgrid is one of the modern Industrial Control Systems (ICS) for the power sector. The system is implemented using industry-standard Virtual Instrumentation (VI) technology. VI refers to using a general-purpose data acquisition system, PC or embedded computing system, a driver software, and a graphical system design software, such as LabVIEW (Laboratory Virtual Instrument Engineering Workbench), to develop a custom-defined VI system for any measurement and control application [1]. It is one of the technologies successfully used for ICS. ICS refers to a broad class of systems that provide the functionality of monitoring, control, and automation in manufacturing, industrial, commercial, and industrial applications. ICS integrates different hardware components and software into a workable system that may be localized, distributed, embedded, PC, or network-based [2].

A general ICS consists of hardware/software controllers, input sensors, and output actuators such as burners, valves, gauges, and motors all work in a tightly integrated manner to perform an automated task. The entire control system is controlled by the data acquisition and control program embedded into the controller or PC with display panels or graphical user interfaces/man-machine interfaces. The GUI of the control system, designed using typical control system software or general-purpose virtual instrumentation software, provides the entire 'view' of the control process allowing the user/operator to see the results and parameter values of control actions graphically and make changes to the specific input parameters of the system [2]. An ICS may be classified as a simple process control system (PCS), distributed control systems (DCS), supervisory control and data acquisition (SCADA) system, network control system (NCS), and many others. According to the industry, each ICS functions differently and uses general or typical devices and protocols. ICS are employed in almost all industrial applications and sectors such as manufacturing, transportation, energy, robotics, water treatment, critical infrastructure, automatic machines or measuring devices, domestic and commercial appliances, etc. [3].

Smart Microgrid

The use of electricity has become an essential part of life at the domestic, industrial, and commercial levels. Today, by and large, everything is utilizing electricity and not only AC power but also DC power. Low voltage DC networks are on the rise with the requirement of DC voltage for a large number of consumer products and sensitive electronic devices such as computers, laptops, LED lighting systems/tubes, etc., appliances with power converters, electrical vehicle charging, battery charging, and upcoming of USB based electrical sockets [4]. Over the past few years, electricity demand has been increasing at a fast rate and it is estimated that power consumption in India is likely to increase from 1160.1 TWh (Terawatt Hours) in 2016 to 1,894.7 TWh in 2022 and up to 2.3 times over the next 20 years [5]. Soon, traditional AC grids may not be able to meet the ever-increasing demand for electricity optimally for different sectors of society.

To overcome some of the limitations of traditional grids in meeting the demand of electricity efficiently to remote, inaccessible locations, the concept of Smart Grid technology has emerged. A smart grid is an intelligent system that improves the power system by integrating renewable green energy resources to provide an automatic flow of electricity to meet consumers' demands more efficiently, economically, and safely with better power quality. The smart grid is the 'computerizing' of the electric utility grid and promotes the use of sustainable energy resources in high demand. In a smart grid, the focus is on the use of distributed, sustainable, and renewable energy systems such as solar/wind/fuel cells and integrating them with the existing grids (ON grid) or using these in isolation (OFF grid) for efficient power generation and distribution to meet the energy demand of variable loads. Generally, the smart grid provides electricity where it is consumed, which can easily be met by renewable green energy resources, thereby reducing air pollution, electricity generation, transmission, and distribution expenditure, along with a decrease in electricity bills. A vital feature of a smart grid is the industrial automation technology that provides utility adjustment and control of each device or millions of devices installed in distributed renewable resources from a central location. Around the world, the share of new electricity access supplied by renewables will nearly double to 60%, up from 34% over the past five years. This pattern is even more extreme in India, where the share of new electricity from renewable will triple to 60% [6, 7].

A more realistic and consumer-friendly approach to the smart grid is microgrid or still on a smaller scale nano-grid or pico-grid. A smart microgrid is the foundation of a perfect power system. It is a modern, small-scale version of today's centralized electricity system. Figure 8.1 shows the conceptual model of a smart microgrid that generates, distributes, and regulates the flow of electricity to consumers by integrating renewable

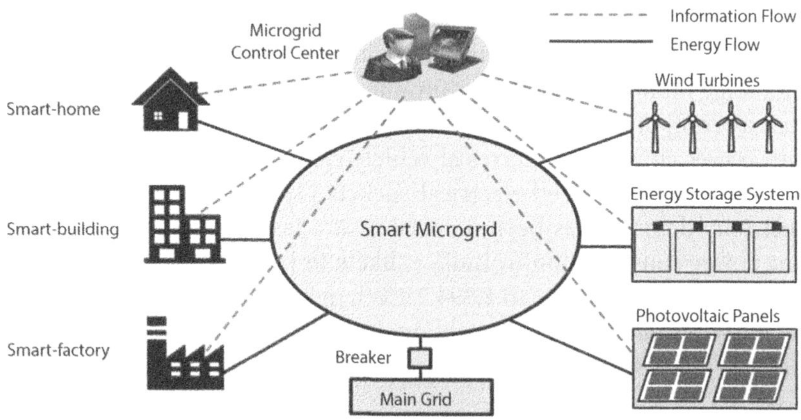

Figure 8.1 Conceptual model of smart microgrid.

resources of energy with or without the utility grid locally at specific locations. All the components of the microgrid are combined to form a single control system. There is a seamless flow of information and electrical energy within the microgrid components [8]. A microgrid can also be networked with one another and the central grid to increase reliability, capacity, and efficiency. However, unlike a centralized grid, a microgrid is not necessarily owned or run by a utility company. A microgrid is built, owned, and operated by a corporation, community, neighbourhood, hospital, university, individual, or any other entity with legal authority over their power infrastructure (i.e., power lines, generation source, meters, etc.). This local control, which allows for private investment in the system and their relatively small size, makes microgrid hotbeds for entrepreneurial innovation. In addition, because a microgrid is built to optimal scale for its users, whether for a single building, home, or community, one can thus establish a network of microgrids to provide tailored power service at the best price to each of the distributed consumers [9].

A microgrid features the latest innovative technology that increases reliability and efficiency of power usage and creates islands of sustainable energy within the larger grid. There is significant variability in the design of a microgrid. It depends upon many factors such as type of power generation capabilities (AC/DC/hybrid), types of renewable energy resources used such as solar/wind/fuel-cell or hybrid, grid connectivity (ON-grid/OFF-grid), and application demands. Among the modern techniques of sustainable or green energy production, solar-based technology is one of the most advanced and promising methods. Solar systems are well suited and easily deployable at the maximum of the locations where sunlight is abundant. Further, to increase the utilization of renewable energy sources for domestic purposes, especially in the rural areas, the OFF-grid power system is becoming popular and is required to be installed to meet community power needs. Many models of microgrids using fixed solar setups on a roof or standalone device have come up. For distributed generation (DG), grid-tied photovoltaic-wind hybrid systems with centralized battery backup are also used to mitigate the variability of the renewable energy resource [4, 10].

Proposed System Overview

A typical hardware and software infrastructure is required with innovative metering capabilities integrated with communication and networking technologies to realize a power-efficient microgrid system. Motivated to re-design some of the crucial subsystems of a microgrid, this work discusses the development of a virtual instrumentation-based advance metering infrastructure model of a microgrid. The proposed system uses modern industrial control strategies for real-time efficient power management of a small-scale DC microgrid in a more informative, user-friendly, and cost-effective manner.

Compared to the related works discussed in section 8.2, this proposed work concerns the design, development, and field-testing of a novel and reliable DC microgrid system that leverages industry-standard virtual instrumentation and wireless data communication technologies. The proposed approach integrates multiple energy resources such as a solar, battery, and DC grid with the Wi-Fi-based embedded hardware and software power management controller with a soft metering panel for efficient power distribution to DC loads.

The proposed system wirelessly transmits the microgrid data to a remote PC at a distance of 100-200m. A dedicated virtual instrument software with GUI manages the entire operation of the microgrid with data logging. The panel controls the operation of the microgrid. It monitors all its vital parameters such as load energy consumption, utilization of power resources, microgrid operating status, and solar temperature alarm. This customizable microgrid system can quickly generate and manage power distribution to DC loads locally to homes, rooms, shops, rural areas, etc.

The chapter is well structured and organized. Section 8.2 discusses the survey of the related works in the area of the microgrid. The design architecture, hardware, and software components of the proposed system are explained in Section 8.3. Section 8.4 discusses the experimental work, results, and performance analysis of the developed prototype of the microgrid system. Section 8.5 summarizes the conclusion along with the future scope of the work.

8.2 Related Work

Over the last decade, there has been tremendous growth in smart microgrid technology that empowers consumers and stimulates future electricity innovations and entrepreneurial activities [6]. Many models, technical design issues, probable solutions, and applications using renewable energy resources in microgrids have been proposed by many authors in the literature. This section provides a systematic overview of sample related works in solar-based DC microgrids covering research ideas to commercialization and implementation at the national level.

Solar energy is eco-friendly and sustainable energy available on earth in abundance. The use of solar power to drive electronic devices has been explored in many products such as solar watches, solar cookers, heaters, power harvesting for small sensor nodes, unmanned aerial vehicles, satellite systems, and the latest, electric vehicles. A microgrid with an appropriate converter and inverter supplies power to DC and AC loads [9]. The concept of integration of renewable energy resources with traditional grids in the ON-grid or OFF-grid mode to overcome energy crisis problems has gained momentum over the years [10]. The realization of solar-powered microgrids requires typical hardware components, application software, power control strategies, etc., as many authors discussed in the simulation environment. The authors have taken modeling and simulation studies on different aspects of microgrid such as a charge controller for a PV module [11], a solar-powered street-lighting system [12], DC solar power for rural Indian homes [13], and an autonomous coordinated control strategy for the DC microgrid [14]. Many authors have implemented various components of a microgrid system for innovative monitoring applications. These include the design of a graphical user interface to monitor smart grid electricity [15], use of the Arduino embedded platform with the LabVIEW and web-browser for solar power energy consumption by loads [16], laboratory testbed for a DC microgrid for a college in New York [17], better energy utilization in DC nanogrid with mobile activated pre-paid billing [18], and many more [19].

A more thoughtful way of power management in India is using a smart microgrid [20]. The government is diversifying its energy generation capabilities at the national level using renewable resources to meet ever-increasing demand at a lower cost. However, there are many pitfalls in various technical and non-technical issues of the Indian power market and power systems, characterized by the poor demand-side management and response for lack of proper infrastructure and awareness as reviewed in the paper [20]. Decentralized renewable energy systems (DRES) or distributed OFF/ON microgrids are designed to meet the specific power needs of different populations on various scales. Therefore, a microgrid is considered a viable solution for all economic segments living in diversifying locations, according to the press report released by the Government of India [21]. India's theoretically calculated solar energy incidence on its land area was about 22 GW per year in March 2018. The government has revised the target of grid-connected solar energy capacities from 20 GW to 100 GW by 2022. The off-grid solar power plant that can easily be installed locally at a lower cost can be used for rural electrification. A smart DC rural microgrid describes a low voltage DC distribution network for rural electrification within an intelligent grid concept [10]. In Rajasthan during FY2016-17, 91 villages were electrified with a stand-alone solar system and over 6,200 households have received a 100W solar home lighting system. India has sold or distributed about 1.2 million solar home-lighting systems and 3.2 million solar lanterns and has been ranked the top Asian market for solar off-grid products [21]. Many companies have recently come up in India to provide solar-based rural electrification and the establishment of 300 microgrids with a renewable energy generation capacity of 15MW over the next four years [22].

8.3 Proposed System

8.3.1 Microgrid Architecture

Figure 8.2 shows a typical proposed metering infrastructure model architecture for automatic power management of a small-scale smart DC microgrid. The system, based upon virtual instrumentation [1] and wireless networking technology [23], integrates multi-power sources (solar power, battery, and grid), hardware components (sensors, battery charge controller, switch control unit, and DC loads), wireless data acquisition (DAQ) real-time embedded controllers, and microgrid power management software for reliable power distribution and management to DC loads. Both

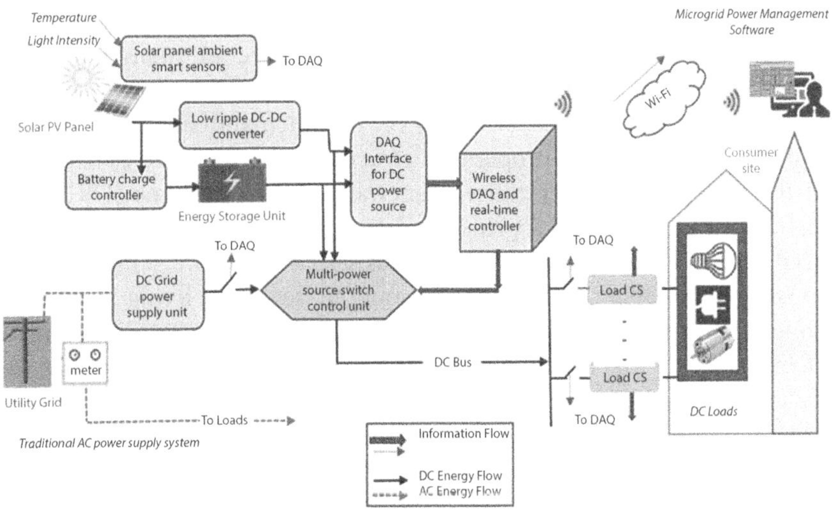

Figure 8.2 Typical architecture of proposed virtual instrumentation metering infrastructure for automatic power management of small-scale smart DC microgrid.

information and electric energy flow in the system in a seamless manner. Electrical signals of power sources and ambient environmental signals of solar panels are interfaced to the embedded controller that sends control signals to multi-power source switches to drive the loads using the most optimum available power resource. The embedded control program acquires, pre-processes, and wirelessly transmits the data over the Wi-Fi network to the remote server for data collection, analysis storage, and display.

The dedicated microgrid power management software with a remote user interface controls the operation of the microgrid. It automatically selects the best available power resource to drive loads to optimize the utilization of solar energy. The software has different programmed modules to define all the functions and features of the system and a versatile virtual microgrid control panel for user interaction. The control panel has soft digital meters to display the operational parameters of the microgrid at the user site.

8.3.2 Microgrid Hardware

Metering infrastructure for the microgrid consists of several hardware components such as distributed multi-power resources, electrical processing interfacing circuits, power switch control unit, DC load with current sensors, and switches. The sensors, switches, and actuators are interfaced to the microgrid controller, a Wi-Fi-enabled reconfigurable data acquisition and real-time processor (DAQ). The important hardware components are explained as follows:

(i) Solar Photovoltaic Panel and Interfacing Circuit: In this microgrid setup, a 20W polycrystalline solar photovoltaic panel is installed in the OFF-grid mode. The solar panel charges the battery and directly feeds power to the load instead of the DC grid. Generally, for economical installation of solar panels, polycrystalline panels with reasonable efficiency (12 to 17%) are preferred against monocrystalline which have a higher efficiency (18-20%) and cost. Thin-film solar panels are gaining acceptance in low-cost, flexible installations but offer very low efficiency (12%) [23].

The output voltage of the solar panel that varies with light intensity is pre-processed using low ripple DC-DC converter LM2596. The converter has an in-build filter that provides adjustable regulated solar voltage to charge the battery sufficiently and drive the loads [24]. Further, the regulated solar panel voltage is interfaced to the data acquisition (DAQ) embedded controller using a voltage attenuator interfacing circuit (Figure 8.5). The circuit uses a potential divider arrangement of two high-power and low-value resistors of 15Ω and 10Ω. The resistor values are calculated using a linear mapping attenuator algorithm that converts solar panel rated voltage in the range 0-25V to a DAQ voltage range of 0-10V without an increase in the power dissipation of resistors. The operational efficiency of a solar panel is also affected by ambient environmental parameters such as excess temperature rise, dust, and shading that eventually deteriorate the panel health and needs consideration, as discussed in this work.

(ii) Electric Storage, Charge Controller, and Interfacing Circuit: An electric storage element such as the rechargeable battery (12V, 4.5Ah) is used as a power backup. The battery voltage is used in case of emergency when solar power is unable to meet the load demand. A lead-acid battery is preferred over lithium-ion or nickel-cadmium for small-scale microgrid applications because of its low cost, small size, low discharge depth, and operation reliability.

The battery is continuously charged by a solar panel using either buck type or a more efficient maximum power point tracking charge controller [25]. Figure 8.3 shows the circuit diagram of an LM317 (three-terminal voltage regulator) based battery charge controller using a transistor and Zener diode switching circuit to charge the battery and prevent its overcharging efficiently. Further, to measure the electrical parameters of the battery, the battery output voltage is interfaced with the DAQ device

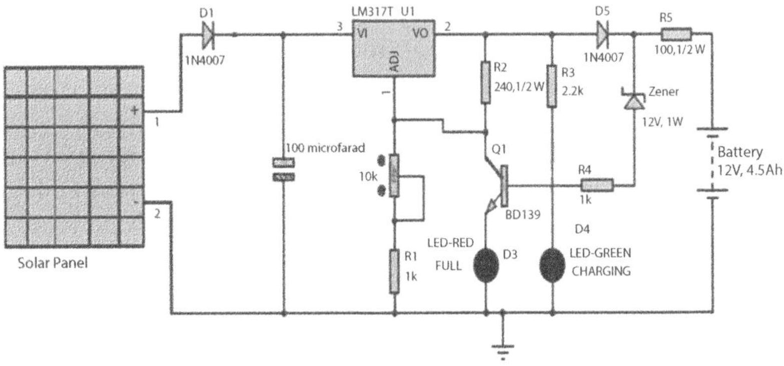

Figure 8.3 Circuit diagram of automatic battery charge controller.

through the attenuator interfacing circuit (Figure 8.5). The attenuator consists of a voltage divider network of high power and low-value resistors of 15Ω and 10Ω resistors. This circuit maps the voltage range of the battery (0-12V) to the DAQ voltage range of 0-10V.

(iii) DC Grid Supply Circuit and Interface: Figure 8.4 shows the circuit diagram of the DC grid voltage regulator. The single-phase AC voltage (230V) available from the utility grid is converted to a 15V regulated DC supply known as a DC grid voltage using a step-down transformer (15-0-15V), a rectifier (diode IN4007 and a capacitor (1000μF, 50V)), and a regulator (IC 7815). Rectified DC voltage provides power to heavy DC loads for which solar energy is not sufficient. The availability of the DC grid supply is checked through the status of the programmable push-button switch that is interfaced with the DAQ controller (Figure 8.5).

(iv) Solar Panel Ambient Smart Sensors: This module uses smart linear, miniature sensors with a direct analog interface to measure ambient

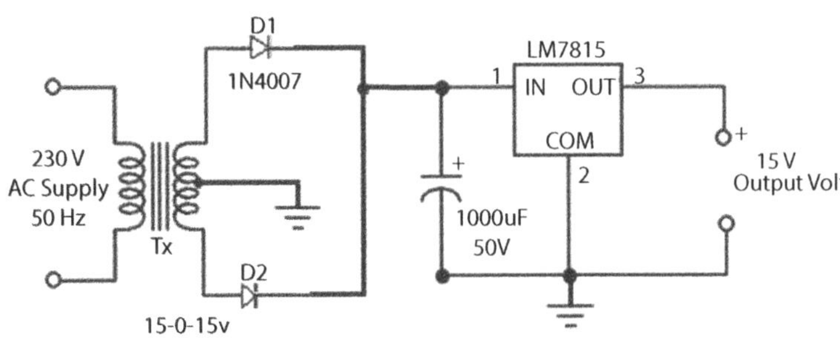

Figure 8.4 Circuit diagram of 15V regulated DC supply voltage grid supply.

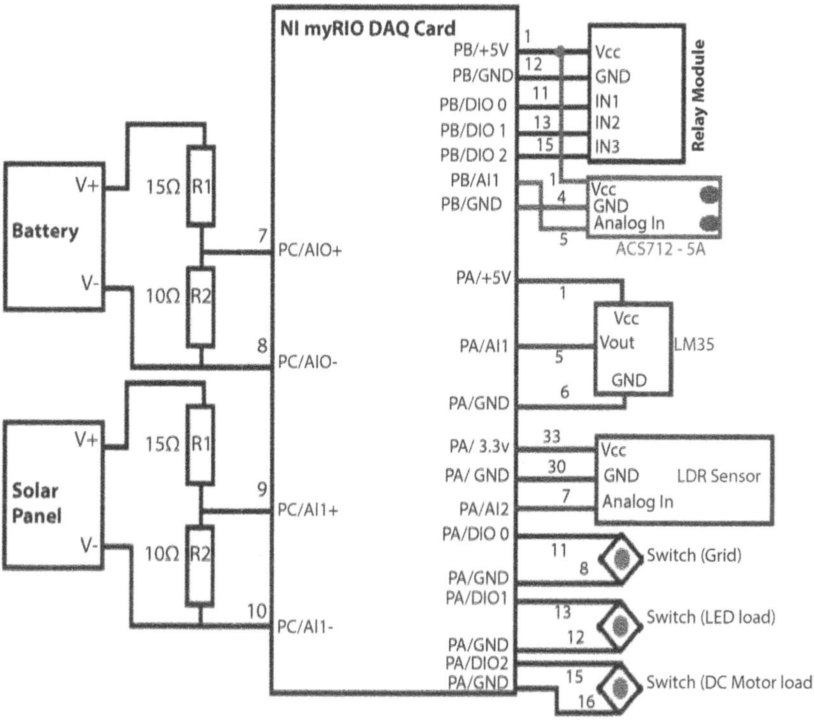

Figure 8.5 Interfacing of NI-MyRIO to microgrid hardware components.

temperature and light intensity falling on the solar panel. Measurement and analysis of these environmental parameters indicate the health and efficiency levels of the solar panel to initiate appropriate alarm messages for the proper functioning of the solar panel. For ambient temperature measurement, LM35, a smart solid-state analog temperature sensor, is used. It is a miniature transistor-like sensor with three terminals (Vcc, Gnd, Vout) conveniently connected underneath the solar panel and interfaced to the onboard data acquisition module (Figure 8.5). This sensor offers good linearity in the temperature range of -40 to 110°C with a sensitivity of 10mV/°C. It requires a low excitation voltage of 3V and has a very low power consumption of 6μA [25].

Ambient light intensity falling on the solar panel is measured using a light-dependent resistor (LDR) fixed on the solar panel. When the light falls on cadmium sulphide (CdS) LDR, its resistance decreases nonlinearly and varies from a dark resistance value of 15KΩ to 500Ω under full light conditions. Resistance variation is further converted to voltage in the range 0-5V using an on-board voltage divider network. The output voltage of the 3-pin light intensity module is interfaced to DAQ, providing 5V excitation to the sensor (Figure 8.5). Using the predefined calibration chart of the LDR, the measured value of voltage is converted to light intensity in lux or W/m^2. In the proposed model, light intensity measurement is used to draw and test the characteristics of the solar panel.

(v) Multi-Power Source Switch Control Unit: For automatic selection and switching to the most appropriate power resources available on the microgrid for the given load, a multi-power source switch control unit is designed and used. It consists of a software-controllable 5V, 4-channel optocoupler relay module interfaced to DAQ, power sources, and load, as shown in Figure 8.5. The control terminal of each relay unit is interfaced to a typical digital line of DAQ and one of the coil terminals of each relay receives standard 5V excitation given by DAQ. Each relay's normally open (NO) terminal is connected to a power source to drive parallel connected loads. Depending upon the microgrid power controller's digital control signals, the relays seamlessly switch among different power resources and manage power to the load automatically and reliably. Using a relay with an opto-transmitter-receiver pair provides sound isolation of high power electrical circuits from a low power control circuit. It has a noiseless and fast switching operation suited for the application.

(vi) Data Acquisition and Real-time Controller: The proposed system uses NI-MyRIO [26] from National Instruments as a typical data acquisition (DAQ) module with a modular I/O, real-time embedded controller, and wireless communication unit. MyRIO is a LabVIEW programmable device configured and controlled through an embedded application program in the processor. The DAQ works like a typical SCADA (supervisory control and data acquisition) system with Wi-Fi communication capabilities for the microgrid. The device is programmed in an embedded LabVIEW application named 'microgrid power management and metering software' to control the operation of the microgrid and execute power management and data visualization functions in real-time [26].

Figure 8.5 shows the circuit diagram to interface microgrid signals to MyRIO. For power management and information processing, all the sensors, switches, and actuator signals of the microgrid are interfaced to analog and digital channels of MyRIO. The voltage output signals from sensors, solar power, and battery are connected to analog input channels. Digital signals from switches are connected to digital input lines of the DAQ and digital control signals available on output lines of the DAQ are connected to the relay control terminals. The interfacing circuits of smart sensors are also powered from the excitation signals available on DAQ without loading the signal line. Apart from performing the standalone data acquisition and control operation, MyRIO also communicates the data to the remote server wirelessly over the Wi-Fi network (802.11b, g, n).

(vii) Load Current Sensor: A smart current sensor based on Hall effect principle ACS712 5A is used to measure the total load current. It has two power terminals to connect load and three terminals for providing output voltage, excitation signal, and common ground. In response to the current variation, the sensor provides an analog voltage proportional to the current. The sensor requires an excitation signal of 5V and has a good sensitivity of 185mV/A. Figure 8.5 shows the interfacing of the current sensor to DAQ and load [27].

(viii) Variable DC Load and Switches: Many devices and appliances work well on DC voltage. Commonly used DC loads in consumer applications such as LED light and DC motor are also used in this system. A typical surface-mountable LED light (10W, 10V) and variable speed permanent magnet DC motor (18500-22000 RPM, 14.4V) are used as a light and heavy load, respectively, for the microgrid. As shown in Figure 8.5, the loads are connected in parallel to the microgrid supply through the respective push button switches and current sensor. Whenever there is a change in load from low to heavy power or vice versa, the power switch control algorithm senses the change in the current and issues signals to the switching circuit to disconnect solar energy and connect grid power to the load. This way, the microgrid automatically manages the flow of electric power from the best available power supply based upon load.

(ix) Microgrid Server: A modern Windows-based desktop PC with Wi-Fi connectivity is used as the remote server to monitor the power management operations on the microgrid control panel. The server executes the LabVIEW program to communicate with the microgrid controller over the Wi-Fi network and display all vital parameters on the control panel user interface. The server is also used for logging microgrid data generated from MyRIO in the system files.

8.4 Microgrid Power Management and Metering Software

Figure 8.6 shows the functional design model of the application program for a microgrid named 'microgrid power management and metering' software

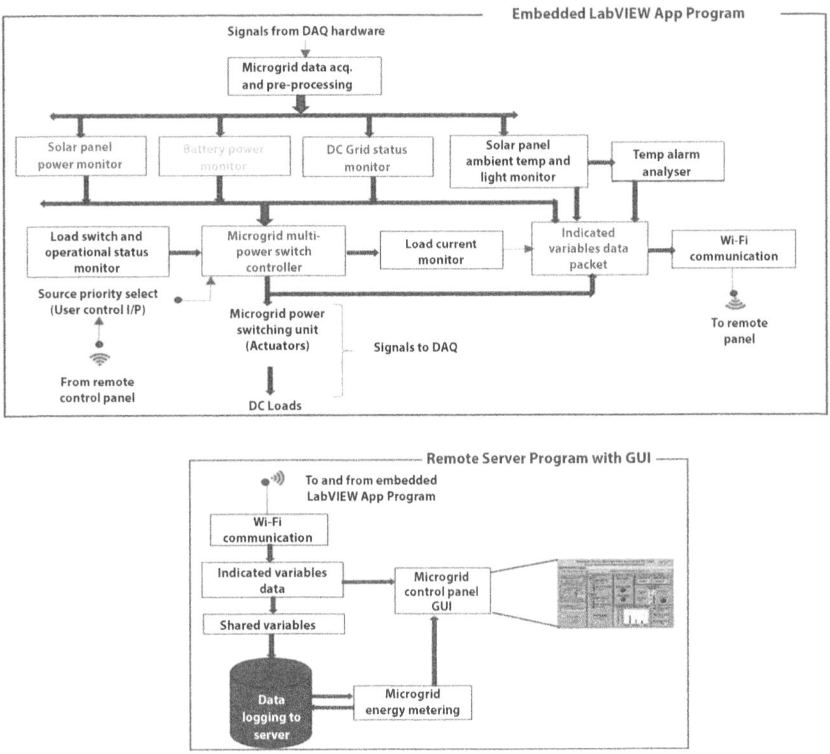

Figure 8.6 Functional design model of microgrid power management and metering software (μG-PMMS).

(μG-PMMS). The software, implemented on the virtual instrumentation platform LabVIEW (Laboratory Virtual Instrument Engineering Workbench), defines the overall functionality and features of the proposed system. The application software has distributed architecture with two virtual instruments (VI): the embedded VI program runs on a MyRIO processor [26] and a microgrid server program with a user interface that runs on the remote PC.

The software defines the overall functionality and features of the proposed system. It has various functional modules integrated to perform data acquisition and analysis tasks, supervisory power control and management, and wireless communication of measured data to the server program for data analysis, logging, and display. The LabVIEW software code is implemented using a block diagram and data flow programming methodology that supports concurrent and serial operation with automatic memory management [1, 28]. The functionality of each module is explained below:

(i) Microgrid Data Acquisition and Pre-Processing: This module is programmed to configure analog and digital I/O channels for data acquisition and processing so that the signals of the microgrid are reliably monitored. Using a continuous data acquisition operation with a 10-point averaging filter, an averaged value of 10 samples of each analog signal is acquired and updated at the acquisition rate of 1s until the STOP button is pressed. Averaging with continuous data acquisition mode improves reliability with which data is measured and reported. The measured value of kth voltage signal at i^{th} time instant ($_{Ski}$) is given by Equation (8.1).

$$S_{ki} = \sum_{i=1}^{j=10} x_{kj} / 10 \tag{8.1}$$

where x_{kj} is the jth sample value of the kth voltage signal acquired at the time instant of sampling.

As discussed in subsequent sections, each of the measured values of voltage signal X_{ki} is further converted into useful information using suitable signal conversion functions.

(ii) Solar-Panel Power Monitor: This module measures the actual voltage and power delivered to the microgrid from the solar panel at the time instant. The DAQ voltage (Vsp_daq) is converted to the actual voltage (Vsp) of the solar panel using the inverse voltage attenuator formula as given by Equation (8.2).

$$Vsp = Vsp\_daq\ (1 + R_2/R_1) \tag{8.2}$$

The power output of this solar panel is measured by Equation (8.3), where the current (I_L) flowing through the interfacing circuit is given by Equation (8.4).

$$P = Vsp * (I_L) \tag{8.3}$$

$$I_L = (Vsp/ (R_1 + R_2)) \tag{8.4}$$

where $R_1 = 15\Omega$, $R_2 = 10\Omega$

(ii) Battery Power Monitor: This module measures the actual voltage stored in the rechargeable battery at a measured time instant to ascertain its power delivering capabilities. The battery DAQ voltage (V_batt_daq) at the time of acquisition is converted into actual battery voltage (V_batt) using an inverse voltage attenuator formula as given by Equation (8.5).

$$V\_batt = V\_batt\_daq\ (1 + R_4/R_3) \tag{8.5}$$

where $R_3 = 15\ \Omega$, $R_4 = 10\Omega$

(iii) Solar Panel Ambient Monitor and Temperature Alert: This module measures the associated health parameters of the solar panel, such as temperature and light intensity, and issues a temperature alarm. The measured voltage of temperature sensor (V_{temp}) in volts is converted into a temperature reading ($T_{ambient}$) in °C using the inverse sensitivity value (0.1°C/mV) of an LM35 sensor, as given by Equation (8.6). Similarly, the measured voltage of the LDR sensor (V_{LDR}) in volts is converted into the resistance value of LDR (R_{LDR}) using a resistive voltage divider circuit, as given by Equation (8.7).

$$T_{ambient} = 100 * V_{temp} \tag{8.6}$$

$$R_{LDR} = ((V_{LDR})^* \ Ro)/(3.3 - Ro) \tag{8.7}$$

where R = 10kohm.

The resistance value of LDR is converted to light intensity values using the calibration look-up table.

The temperature alarm algorithm statistically analyses the temperature profile of the solar panel over the last 15 minutes. If the average 15-minute ambient temperature exceeds the high-temperature limit of 40 ±1°C, then the solar panel temperature alarm is activated. Also, a corresponding alarm message is generated on the virtual panel. The temperature alarm limit is selected from the temperature performance characteristic of the solar panel, where the panel's efficiency decreases appreciably.

(iv) DC Grid Status Monitor: This module checks the availability of onsite rectified DC grid voltage by reading the status of the switch on digital line DIO0, as given by the algorithm below.

Read DIO0

If signal HIGH (≥4.5V) --> Switch Closed --> DC Grid power supply available

Otherwise, Switch Open --> Supply OFF.

(v) Load Switch Status and Current Monitor: This module detects the ON/OFF status of load switches connecting light/heavy DC loads to infer the operational state of load (light load, heavy load, Off). The status of the switch is detected by reading the signals on respective digital input line/s (Signal_DIO<line>) and comparing it with the pre-set high voltage level of 4.5V, as given by the algorithm below.

Read DIO <line>
If Signal_DIO<line> ≥ 4.5V *Switch ON*
Otherwise → *Switch OFF*

Information regarding the operational status of the load is used as one of the parameters to operate a microgrid power-switch control unit. When the load switch changes to a heavy load, the grid power is ON, otherwise, the load is powered using solar or battery.

The current monitor program finds the power consumed by the loads using the load current sensor reading. The measured value of current sensor voltage (V_CS) in mV is converted into load current (I_load) in mA using the inverse sensitivity value (S^{-1}) of the current sensor and is given by Equation (8.8).

$$I\_load = S^{-1}\,(V\_CS) \times 100 \tag{8.8}$$

where S is 185mV/A.

(vi) Microgrid Multi-Power Switch Controller: This module manages DC power distribution to loads from any one of the multiple power sources of the microgrid smoothly and efficiently. Figure 8.7 shows the flowchart of the microgrid multi-power switch controller for efficient power

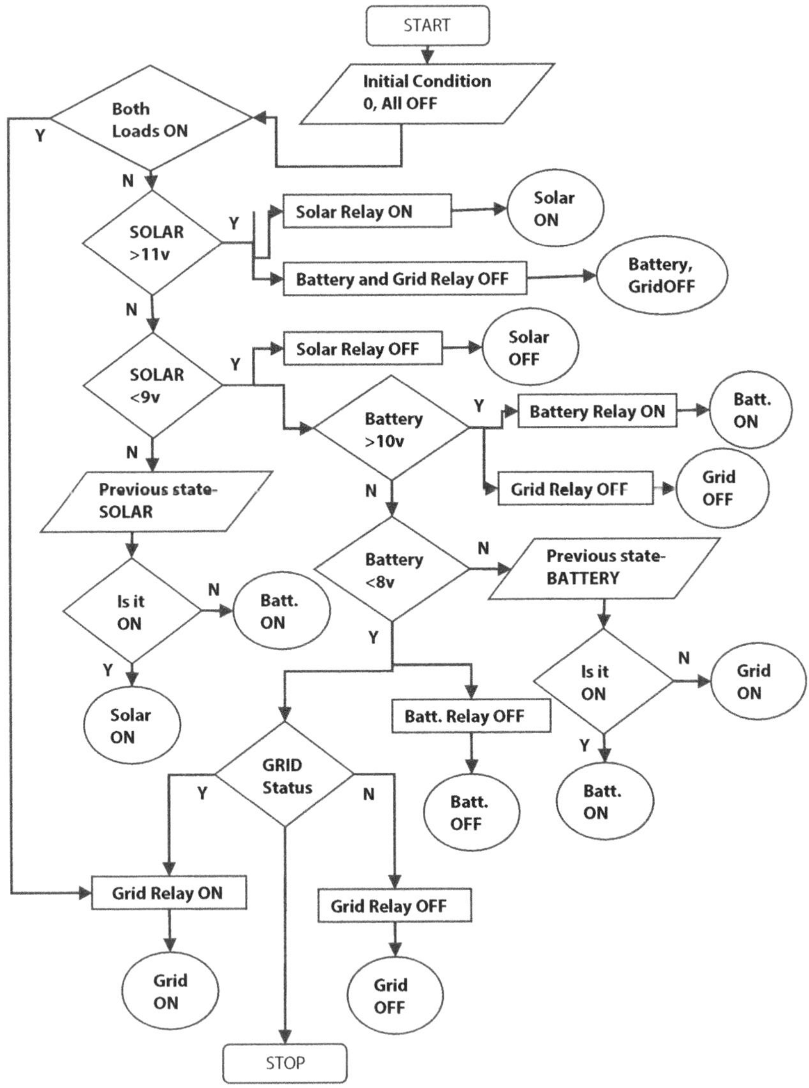

Figure 8.7 Flowchart of microgrid multi-power controller for efficient power management.

management. The algorithm incorporates supervisory control strategies with a finite state machine model to balance the demand and supply of electrical power [28, 29].

The microgrid has a finite fixed number of operational states designated as Solar ON, Battery ON, DC Grid ON, or all OFF. At any time, the microgrid remains in one of the operating modes, indicating its working status. Depending upon the availability of input power from different sources, supervising the selection of the type of the load, load power requirement, and user preferences of power sources, the controller selects the particular power resource to operate the load. The controller then issues appropriate digital control signals to all relays of the power switch circuit to connect or disconnect the power resources to meet load requirements using a dead-band switching algorithm [30].

Further, to select the prioritization of power resources, two modes of preference are provided such as Mode1: 'Solar-Battery-Grid' and Mode 2: 'Solar- Grid- Battery', which incorporate a practical and helpful way to prioritize the power resources in the microgrid as per the user requirement. In both these modes, the use of solar power is given the highest priority. Mode 1 is the most economical as the DC grid, which involves more electrical charges overhead and is least preferred. However, to save the battery for critical times, Mode 2 is preferred. The aim is to optimize the use of solar power and save the battery for critical times.

The entire operation of the microgrid to measure-check-switch the most optimum power source is automatically performed at each time instant of measurement with a negligible time lag and relay chattering. The microgrid changes its operational state according to the priority mode selected whenever the current power source cannot meet the load requirement. For example, in preference Mode1, if the load selected is 'light', the controller checks the availability of solar power. If it is available to drive the load, solar is switched ON and connected to the load. Otherwise, the controller checks the battery power source. If battery power is available, it connects battery power to drive load (Battery ON). If both the power sources are unavailable, the microgrid connects the electrical power from the available DC grid (DC grid ON). However, for a heavy DC load, the microgrid directly switches to a DC grid as a solar/battery power used in this prototype model cannot drive the heavy load. At each time instant, load power consumption from a particular connected source is also measured and logged in the file system.

(vii) DC Microgrid Control Panel: The user interface of μG-PMM software is the control panel of the microgrid at the remote server. The panel, designed on LabVIEW [29], has a systematic and logical layout with menu-driven controls and multi-display objects. The virtual panel allows the user to operate the microgrid, select user inputs, log data to the server file system, and display vital parameters on digital and graphical indicators. The important functions of the panel are described below:

- Remote Communication with Microgrid Controller: The GUI communicates with embedded LabVIEW source code in MyRIO (microgrid controller) over the Wi-Fi network to share microgrid real-time data remotely. Wi-Fi is a standard wireless technology, readily available in all modern PC, data acquisition devices, and embedded controllers to support the development of wireless local area network (LAN) applications [31, 32]. Acquisition data of microgrid and user selection of the input, such as priority mode of power resources, are wirelessly communicated among the server and embedded program. The program execution results are also shared from the MyRIO to the remote server for analysis and display on GUI.
- Data Logging: Data related to the real-time operation of the microgrid at each instant as communicated from the MyRIO controller is dynamically logged in the server files using shared variables [28]. The logged data is analyzed to display the vital parameters on the GUI.
- Energy Metering and Data Display: This module calculates the total energy consumed by the load using different power resources and displays the percentage utilization of the energy at each instant during the day. At each ith measurement point, the total energy consumed by load using an hth power source (E_{hi}) is given by Equation (8.9) and is logged in a file.

$$E_{hi} = V_i * IL_i * \Delta t \tag{8.9}$$

where V_i is the voltage across the load

I is the current flowing through the load

Δt is the measurement time interval

The total energy consumed by the load using the hth power source is the sum of the incremental energy until the current time and is given by Equation (8.10).

$$E_{Th} = \sum E_{hi} \tag{8.10}$$

for all power sources.

Another essential criterion to analyze the operation of the microgrid is to record the utilization of power resources. Percentage utilization ($\%U_{PS}$) of the energy resource is given by Equation (8.11).

$$\%U_{PS} = (T_{PS}/T_{total})\ 100 \tag{8.11}$$

where T_{PS} is the total time for which the power resource is used and

T_{otal} is the total time for which the microgrid is operative for the day.

Apart from the display of load energy consumption and percentage utilization of power sources on GUI, other vital parameters of the microgrid displayed are absolute date/time, solar voltage, battery voltage, type of load selected, the status of grid power availability, operating status (ON/OFF) of solar, battery and grid, ambient light intensity, ambient temperature, and temperature alarm for the solar panel.

8.5 Experimental Work and Results

The system prototype was developed to demonstrate the operation and analyze the real-time operational performance of the proposed microgrid power management system. The solar panel was installed on the rooftop and is hooked to the microgrid system installed in the laboratory. The system was controlled remotely over a Wi-Fi network through the virtual control panel. Many experiments were conducted under different conditions to see the system response and analyze the results. The section below discusses the results.

(i) Onsite Solar Panel Testing and Characterization: To test the onsite working performance of the 20W polycrystalline solar panel, it was placed outside with no sun tracking and electrical parameters (short circuit current and open-circuit voltage) were measured during the day. Figure 8.8 shows the characteristic curve of the solar panel. With the increase in light intensity from the initial value to the maximum value of 79000 lux during the day, current and power output values increased and then started decreasing as light intensity decreased. The open-circuit voltage remained nearly constant, around 20V, with the variation in the light intensity. Short circuit current increased from 0.18-1.1A with power output varying in the range 3-24W. The maximum values of the parameters were near the rated values. So, it was expected that the solar panel would be able to drive

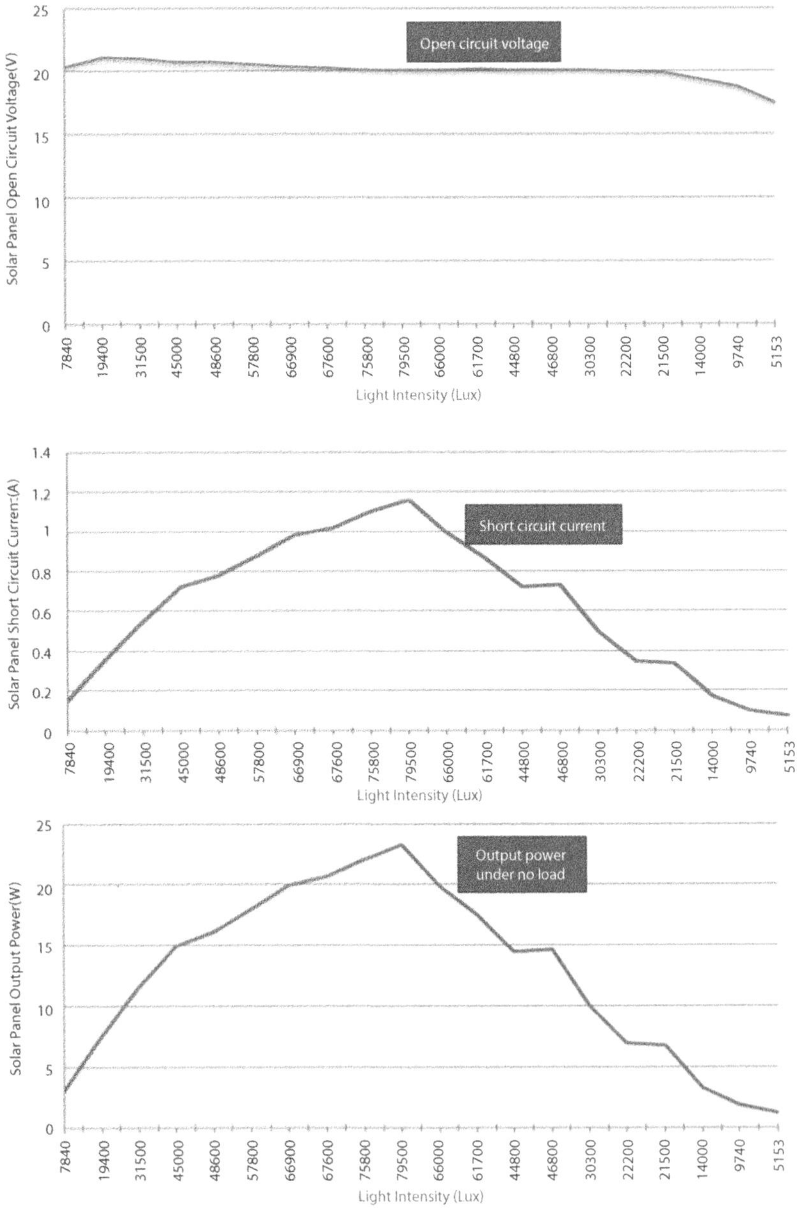

Figure 8.8 Characteristics of 20W solar panel under no load for variations in light intensity.

low-power DC loads with a power requirement of 10-20W, such as LED lights and a small DC fan. Additionally, the solar panel would charge the battery through the charge controller for the considerable time during the days of bright sunlight through the charge controller.

(ii) Real-time Power Management Operation of a Microgrid: To start and visualize the online operation of the microgrid, the system prototype was powered ON. A DC microgrid control panel (GUI) on the remote PC was opened. Prior to the program execution, user input parameters such as priority of the power resources and type of load (light or heavy) were selected. When the RUN button was clicked, the embedded program was executed. At each point of measurement, depending upon the power requirement of loads, availability of power from power resources, and priority selection of the power resources, the microgrid controller issued signals to connect the most optimum energy resource to the microgrid to drive the loads. Figures 8.9 and 8.10 show the results of system operation under different conditions.

Case 1 Solar Panel ON: Figure 8.9(a) shows the screenshot of the DC microgrid control panel displaying the operational status of a microgrid when the load selected was 'light' and power source priority selected was 'Solar, Battery, Grid'. The load status indicator displayed the selection of LED as 'light load'. At an ambient light intensity of 38572.6 lux, the solar output.

During the day, voltage was 15.2V and the current was 0.6A. Battery voltage was high (12.63V) and grid supply was also available (ON). As the solar energy was sufficient to drive the load, the controller connected the solar power to the load and disconnected all the other power sources. The operating status of solar remained ON, whereas the battery and grid remained OFF. The microgrid operational status was 'Solar ON'. Energy consumption of the load until the system was operative using different power sources such as solar and battery, was cumulatively displayed on the respective smart energy meters. Also, the percentage utilization of the power sources was indicated on the bar graph. Solar energy was used for the maximum time (68%), followed by battery (26%). DC grid supply was used for a very short time (6%). As the average ambient temperature crossed the temperature alarm limit of 40°C, the solar panel high-temperature alarm indicator was ON along with the display of the alert message.

Case 2 Battery ON: Figure 8.9(b) shows a screenshot of the DC microgrid control panel displaying the operational status of the microgrid as 'Battery ON'. Light intensity on the solar panel was high, but as the solar panel was partially shaded, its voltage (5V) and current (0.2A) remained

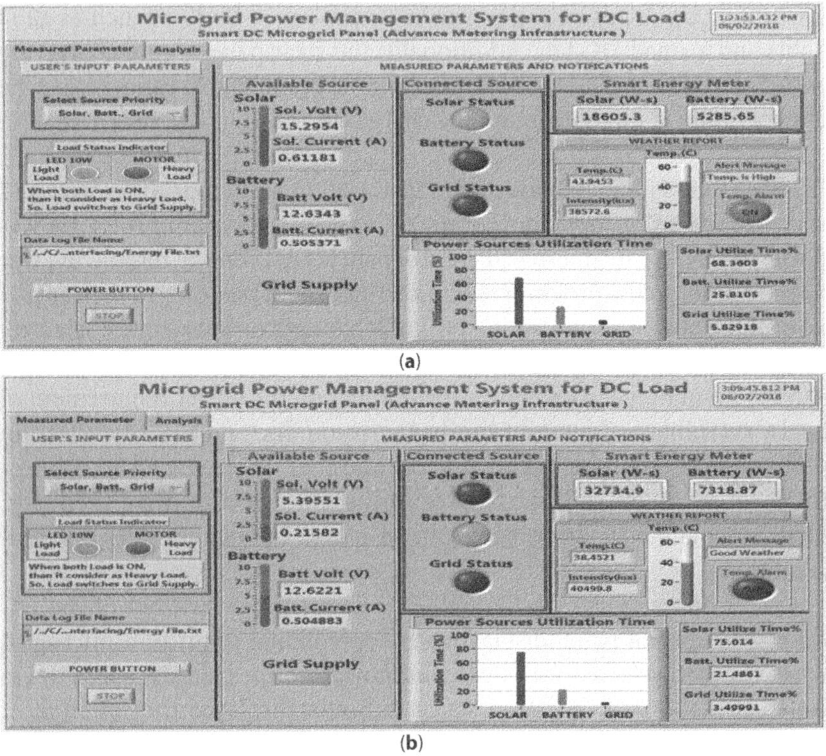

Figure 8.9 Microgrid control panel indicating vital parameters and operational status of microgrid: (a) solar ON; (b) battery ON when solar is not available to manage power distribution to load.

low. Battery voltage (12.6V) and current (0.5A) was sufficient to power the load. The controller issued signals to the actuators to connect the battery to load and disconnect the battery and grid power. Hence, the status of power sources changes, the Battery is ON, and the other power sources are in an OFF state. The temperature alarm for the solar panel is deactivated.

During this time, solar energy was consumed for a maximum of the time (75%), followed by battery (21%) and DC grid supply (3. 4%).

Case 3 Grid ON: Figure 8.10(a) shows a screenshot of the DC microgrid control panel displaying the operational status of the microgrid as the DC grid ON at a time instant. When both the LED light and motor loads were operative, the status of the load changed from light to heavy. Though solar and battery power were available, none of these was sufficient to drive the heavy load. So, a microgrid controller connected the DC grid supply to the load. Grid status was changed to ON and other power sources remained

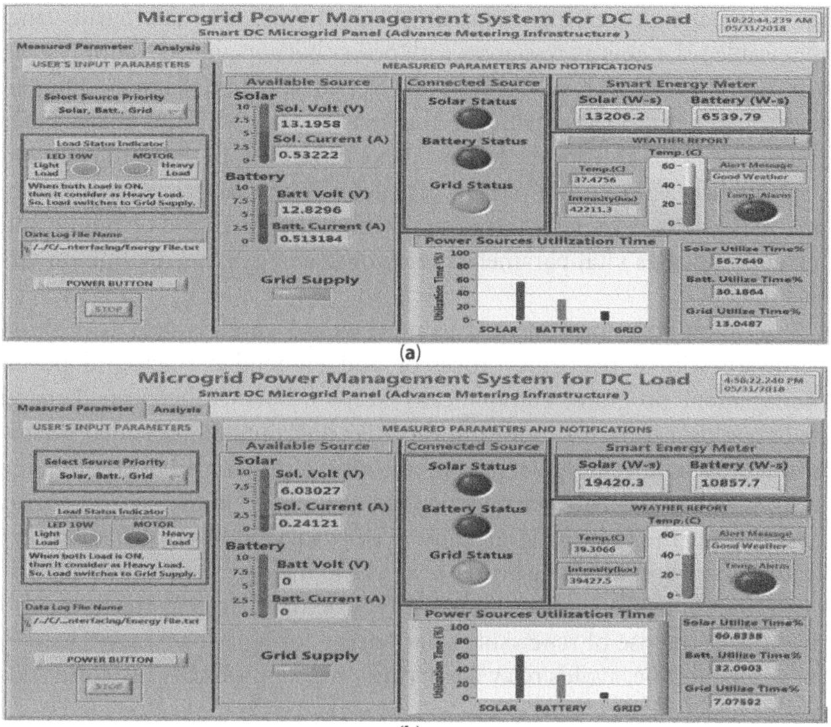

Figure 8.10 Microgrid control panel indicating vital parameters and operational status of microgrid: (a) DC grid ON when solar and battery power are not available for LED light load; (b) DC grid ON when heavy motor load is operated.

OFF. Another situation when the microgrid worked on the DC grid was the non-availability of solar and battery power for the light load, as shown in Figure 8.10(b). In this setup, the light load was operative. Due to shading on the solar panel, solar output power remained low and the battery was temporarily switched OFF. Since both solar and battery power were unavailable, the controller connected the DC grid to the load and disconnected others. The grid status remained ON and the other power sources remained OFF. However, solar power utilization time for the day remained maximum.

8.6 Conclusion

This chapter presents advanced metering infrastructure with integrated hardware and software architecture as an effective and practical solution for automatic power management of small-scale DC microgrid using renewable solar energy resources with power backup and a DC utility grid. The overall system is successfully developed and tested under different real-time energy-load situations. The system can distribute and manage DC power to loads from the best available power resources seamlessly and continuously with a time rate of 1s or less. The microgrid user interface provides access to valuable information about the microgrid operation and instantly updates vital parameters, including energy consumption on virtual meters.

The microgrid, implemented using standard wireless data acquisition hardware with a real-time processor and embedded microgrid power management program, automatically keeps a balance between demand and supply of electrical power. It facilities distributed embedded microgrid control and remote wireless monitoring of microgrid operation on a customized microgrid user interface (control panel) at the consumer site. At any time, the finite state machine supervisory power switch controller connects the best available power resource to the load based upon the prioritization of power resources and the type of load selected. The microgrid maximizes the use of renewable solar energy for light loads, saves battery for a critical time, avoids relay chattering, and conditionally uses grid supply for heavy loads.

The microgrid controller also communicates with the remote control panel (user interface) over a secured Wi-Fi network that updates and displays the real-time information of the microgrid at a fast rate. The panel indicates variations in electrical parameters and operational status (ON/OFF) of the power resources of the microgrid, load status, load energy consumption using different power resources and percentage utilization time of each power resource. The control panel also incorporates safety and test features such as temperature alarm for the solar panel, measurement of light intensity for testing, and characterization of the solar panel.

One of the novel features of the system that allows the user to select the prioritizing mode of the use of power resources of the microgrid makes the system more flexible with a high degree of utility. The system conveniently handles power requirements under different situations, such as preserving battery power for long power cuts or non-availability of renewable power resources, distributing power to OFF–grid rural/isolated areas using renewable solar power with battery backup. Further, the system increases the reliability of power availability at a lower cost with conditional use of the renewable energy resources and the utility grid for different load types and optimizing solar energy for light loads, thus saving grid energy.

The proposed microgrid model with novel functions and features is cost-effective, reliable, and is an efficient power distribution and monitoring system that can easily be scaled for large power networks and capacity. In the future, the system can further be enhanced and upgraded using integrated hardware and software virtual instrument solutions for a hybrid (DC/AC) microgrid well-suited for homes and shops.

References

[1] Gupta, S. & John, J. (2005). Virtual instrumentation using LabVIEW, Tata McGraw-Hill Education India.

[2] Knapp, E. D. & Langill, J. (Dec 2014). Industrial network security: Securing critical infrastructure networks for smart grid, SCADA, and other industrial control systems (2nd. Ed), Syngress.

[3] Rodriguez. E. (2021) Industrial Control Systems (ICS): System Types & Example https://study.com/academy/lesson/industrial-control-systems-ics-system-types-examples.html.

[4] Salomonsson, D., & Sannino, A. (2007). Low-voltage DC distribution system for commercial power systems with sensitive electronic loads. IEEE Trans. Power Del., 22(3), 1620–1627.

[5] Central Electricity Authority. (2016) Draft National Electricity Plan vol. 1, Govt. of India, Ministry of Power.

[6] Lakshminarayana, S. & Anjul. (2014). Smart grid technology and applications. In Proc. of IEEE Conference on Power and Energy Systems: Towards Sustainable Energy. Bangalore, India, 1-6.

[7] BP. (2020) Statistical Review of World Energy (June 2017). BP. Energy Outlook 2017 https://www.bp.com

[8] Siemens. (2011). Microgrids, Siemens AG.

[9] ABB (2015). Microgrids: A Primer for policymakers.

[10] Shadmand, M.B., & Balog, R. S. (2014). Multi-Objective optimization and design of a photovoltaic-wind hybrid system for community smart DC microgrid. IEEE Transaction on Smart Grid, 5 (5), 2636-2643.

[11] Khera. N., Rana, N., Narendiran, S., Sahoo. S., Balamurugan, M., Karthikeyan.s., & Raglend, I.J. (2015). Design of charge controller for solar PV systems. In Proc. of IEEE International Conference on Control, Instrumentation, Communication and Computational Technologies, Kumaracoil, India.

[12] El-Faouri, F.S., Sharaiha, M. & Bargouth, D. (2016). A smart street lighting system using solar energy. In Proc.of IEEE PES Innovative Smart Grid Technologies Conference, Europe

[13] Jhunjhunwala, A., Lolla, A. & Kaur, P. (2016). Solar-dc Microgrid for Indian homes. IEEE Electrification Magazine, 10-16.

[14] Wang, B., Sechilariu, M., & Locment. F. (2012). Intelligent DC microgrid with smart grid communications: Control strategy consideration and design. IEEE Transaction on Smart Grid, 3(4), 2148-2156.
[15] Dang, S., Kakimzhanov, R., Zhang, M., & Gholamzade, A. (2014). Smart grid-oriented graphical user interface design and data processing algorithm proposal based on LabVIEW. In Proc. IEEE 14th International Conference on Environment and Electrical Engineering, Poland.
[16] Haider-e-Karar, Khuwaja, A. A., & Sattar, A. (2015). Solar power remote monitoring and controlling using Arduino. In Proc. of IEEE Conference on Power Generation System and Renewable Energy Technologies, Islamabad, Pakistan.
[17] Salehi, V., Mohamed, A., Mazloomzadeh, A. & Mohammed, O.A. (2012). Laboratory-based smart power system, Part I: Design and system development. IEEE Transactions on Smart Grid, 3(3), 1394-1404.
[18] Ibrahim, M., Chowdhury, S. A., Huda, M.N. & Khan, M.R. (2016). A smart solution for solar DC nano grid integrated with pre-paid billing and mobile application. In Proc. of 3rd International Conference on Electrical Engineering and Information Communication Technology, Bangladesh.
[19] Hamouda, Y. E.M.,Dwedar, S.J.I.(2020). Optimally automated home management for smart grid system using sensor networks: Gaza Strip as a Case Study. Technology and Economics of Smart Grids and Sustainable Energy, 5:16,1-14.
[20] Karmakar A. & Chattaraj. S. (2015). A smarter way of power management in India using smart microgrid: A Review. International Journal of Recent Advances in Engineering and Technology, 3(5), 2347-2812.
[21] Ministry of New and Renewable Energy. (2018). Solar energy press report of ministry of new and renewable energy. Government of India. https://mnre.gov. in/
[22] PV 2ech. (2017). Azure power wins mini and microgrid projects in Jharkhand. India.
[23] Kalogirou S. (2009). Solar Energy Engineering: Processes and Systems. Academic Press. pp 469–517.
[24] On semiconductor. (2008). LM2596, Step-Down Switching Regulator, Semiconductor Components Industries, LLC.
[25] Lokesh Reddy, M., Pawan Kumar, P.J.R., Chandra, S.A.M, Sudhakar Babu T.& Rajasekar, N. (2017) Comparative study on charge controller techniques for solar PV system, Energy Procedia, 117,1070-1077.
[26] National Instrument. (2016). NI MyRIO-1900, User guide and specification.
[27] Texas Instruments. (2019). Linear Integrated Hall-effect sensor ACS712: Texas Instruments datasheet: Dallas, 2X, USA: 1–15.
[28] Bitter, R., Mohiuddin, T. & Nawrocki, M. (2007). LabVIEW Advanced Programming Techniques, CRC Press.
[29] Folia, S. (2011). Practical Applications and Solutions using LabVIEW™ Software, InTech.

[30] Bhanot, S. (2008). Process Control: Principles and Applications, Oxford Press.

[31] Juca, S., Paulo, C.M., Carvalho & Pereira, R. (2014). Wi-fi monitoring embedded system for electrical microgeneration using renewable energy sources. In Proc. of the 8th International Conference on Sensor Technologies and Applications, 21- 26, Portugal.

[32] Pahlavan, K. & Krishnamurthy, P. (2021). Evolution and impact of Wi-Fi technology and applications: A historical perspective, International Journal of Wireless Information Networks, 28, 3–19.

Part 2

CONTROL STRATEGIES FOR PRACTICAL SYSTEMS

9

Execution of a Portable Fuzzy Controller for Speed Regulator Brushless DC Motors

D. Stalin David

Vel Tech Multi Tech Dr. Rangarajan Dr. Sakunthala Engineering College (Autonomous), Chennai, India

Abstract

Brushless direct current motors (BLDCM) are among the most powerful motors available for usage in industries and automobiles. Because of its great performance, this electric motor is widely used in electric automobiles and other electric vehicles (EV). To maintain the rotor's alignment, it is essential to do an analysis of the hall sensors, which complicates the process during runtime. For the BLDCM Motor's speed controller, it is recommended that an Adaptive Fuzzy Optimal Power Control (AFOPC) control be used. The logical implementation of the framework differs from the usual Fuzzy Logical Controller (FLC), which either speeds up or slows down the error depending [1] on whether the device utilizes simply three fundamental laws or a combination of these three laws. The reference current performance of the controller is implemented by the motor process, which makes use of Pulse Width Modulation (PWM) power. Overall, the proposed controller will be computationally light and steady-state offset implementations will have a lesser tolerance than the proposed controller. For high-performance [2, 3] applications, the AFOPC controller includes an FLC controller and it is designed with a constant architecture that picks the controller that triggers [4, 5] the project depending on the altered operating circumstances. In addition, the AFOP Controller is capable of producing PWM for the inverter as well as balancing the speed of the BLDCM in accordance with error variance. The simulation models for the implemented framework were created in the MATLAB 2017 b software environment and the BLDCM's speed control is being tested under a range of load torque variations. MATLAB 2017 b was used to create the simulation models. It is shown by the results of the simulations that the proposed AFOPC controller produces repeatable results under all operational situations.

Email: sdstalindavid707@gmail.com

Vipin Chandra Pal, Suman Lata Tripathi, and Souvik Ganguli (eds.) Industrial Control Systems, (183–202) © 2024 Scrivener Publishing LLC

***Keywords*:** Adaptive fuzzy optimal power control (AFOPC), speed sensor, driver circuit, brushless DC motor (BLDCM)

9.1 Introduction

The creation of an automotive system is a promising and effective methodology for individual and urban transportation because of the fuel energy source exhaustion and societal concerns. Because of the constrained life expectancy of batteries, an electric motor for commercial applications is ordinarily manufactured [6] to upgrade the power efficiency with a decreased volume. The advancement of permanent magnet materials [7] and the idea of present-day converter technologies and some Adaptive Fuzzy Optimal Power Control (AFOPC) inverter control motors, as shown in Figure 9.1, are extensively used in automobile applications. In the current scenario, BLDC motors are one of the best performance and motor types. It is the most popular motor. These engines are suited for safety-critical requests and are mainly used in the automotive and manufacturing sectors. The BLDCM is comparable to DC motors from a modeling perspective. It has a non-linear connection between contemporary and rotation, energy and speed. The operation of BLDC motors is provided by a mechanically commutated [8] conventional brush motor electronic control system instead. Brushed DC motors have much compensation over other predictable motors. BLCM speed control is based on speed and rotation characteristics, high dynamic comeback, high performance and reliability,

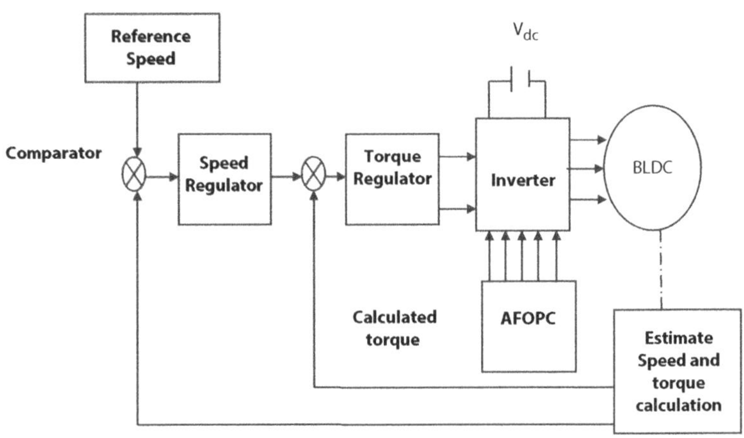

Figure 9.1 Block diagram for ANOC pulse width modulation based inverter.

long working life, noiseless range, high-speed range, and reduced electromagnetic interference.

The BLDCM is similar to DC motors from a modeling viewpoint. It has a non-linear connection between current and rotation, power and speed. BLDC motors commutation is provided with an electronic control system, instead of a mechanically [9] commutated typical brush motor. BLDC motors have many advantages over other conservative motors, such as scrubbed DC motors and induction motors. The speed control of the BLCM is based on features like speed and rotation characteristics, high active response, high efficacy and consistency, long working life, quiet operation, advanced speed range, and reduced Electromagnetic Interference (EMI).

In this setting, modeling, switch strategy, simulation, and parameter tuning of complex process control are executed in the BLDC motor switch system. To obtain AFOPC, the driver system model analysis is required to modify the system parameters used. For the speed control of the BLDCM, the proposed AFOPC controllers must have different hall sensor signals [10, 11], parameter inconsistency, and ambiguity of the torque value; all these parameters are based on the scientific model of the system. The AFOPC based special and fine-tuning parameters [12] are based on the centroid position of the speed system. Based on the position, the Adaptive Fuzzy Optimal Power Control (AFOPC) based controller will produce different iterations [13]. Based on the iteration value, the PWM of the inverter is varied, hence the speed of the BLDCM is stabilized. The benefits of the Adaptive Fuzzy Optimal Power Control (AFOPC) simple and effective control of the oriented scheme using linguistic variables to give BLDCM speed and best stability response under its load variation condition.

The purpose of this work is to examine the speed reaction of BLDC drive systems with Adaptive Fuzzy Optimal Power Control (AFOPC) and fixed/variable load torque controllers. The controller is calculated to control the BLDC drive system's speed based on the speed system's centroid position. The AFOPC controller will generate the appropriate PWM to the power basis inverter which controls BLDC motor speed.

9.2 Related Works

Square-wave phase BLDC motors for greatest torque per ampere control construct a scale that assures it is Proportional Integral-PWM (PI-PWM) driven in the majority of cases. The operation of the PMLDC motor on the fixed plane demonstrates that the control technology's design and analysis are rather straightforward [1]. When suitable parameter modifications are made, the implemented current variation has a satisfactory transient [13] response and zero steady-state error according to the results of the stability study. To correctly modify the speed of the motor without overshooting [14], use the PI controller [2] to make the necessary adjustments. If the non-standard small inductor is comprised of BLDCM asymmetric trapezoidal waves and a three-phase counter-electromotive force, the sensorless method must first perform a position analysis of the non-ideal point detected counter electromotive force transducer [3–5] before proceeding with the sensorless method. To provide more accurate and trustworthy control under dynamic conditions, a novel optimization approach based on the parameter PI controller to regulate the BLDCM speed [6] is suggested in this paper.

It is initially recognized as an alternative to the traditional static model of 5 phase BLDCs in the frequency sphere in this system and an explanation of the implementation of guided sensor control is performed using the sinusoidal internal model [7] as the basis for the guided sensor control implementation.

There is a new rotation reduction control technology [15] available with the system which uses a linear lowered orientation voltage to change an already-existing phase to make adjustments. New motor control technology, according to analysis and test findings, decreases vibrations while simultaneously boosting the performance and maximum speed of the drive motor [8–14]. The low and average power light is driven by an innovative digital Quadral Duty Cycle (QDD -PWM) to depict internal and manufacturing applications using BLDC motor technology using low and average power. The QDDPWM BLDC motor drive for different applications [15] performed adequately when compared to the reference speed and weight circumstances [16].

An Electrical Equivalent Circuit (EEC), such as machine modelling, may be based on a variety of ways to simplify numerous theories, or analysis methods. [16] On the other hand, such complimentary modes provide numerical modes (NM) with more strong solutions. The effect of rotations, as well as reduction and speed control may be achieved by using a tiny DC bus capacitance.

There are a variety of assumptions that may be used to construct a machine model, such as an Electrical Equivalent Circuit (EEC), which can be conducted by a variety of ways that can be used to simplify the analysis methods [17]. In contrast to this, mathematical techniques (AM), such as the finite element approach, provide more confident solutions [16]. The effects of torque ripple reduction and speed control were investigated in the context of a tiny DC bus capacitor. Large capacitors with high performance to provide more accurate and trustworthy control under dynamic conditions, a novel optimization approach based on the parameter PI controller to regulate the BLDCM speed [17]. This is the situation when mapping is used. A neuro-fuzzy controller (NFC) based on a new approach for correcting phase errors in the high-speed range [18–21] and based on a novel way for correcting phase errors in the high-speed range delivers precise commutation points.

9.3 Materials and Methods

The adaptive fuzzy optimal power control (AFOPC) based speed control of the BLDCM system that was used in this study is a block diagram of the system. One of the goals of the system is to simulate the investigation of different control approaches, speed control loops, and algorithms that are utilized in the majority of the control programs [22, 23] that are integrated into this work. According to the present output error and its time derivative, the fuzzy inference system employs the discovered weight value ratio multiplied by the setpoint determined by the fuzzy inference system. Both the overshoot and the rising time are significantly decreased as a result of this technique. The error domain of the system's non-linear proportional [24] gain is shown in the Figure 9.2 below. The linear function of the systems will be generated by the Adaptive Fuzzy Optimal Power Control (AFOPC) controller that has been built. In accordance with the linear function value, the speed of the brushless DC motor is stabilized as shown in the below Figure 9.2.

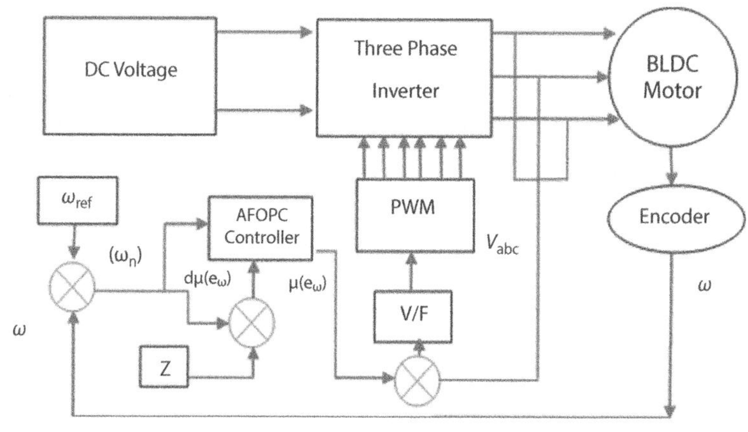

Figure 9.2 Proposed block diagram for BLDC motor.

9.3.1 Scientific Model for BLDC Motor

The scientific model allows the study of BLDC motor behavior under different speed conditions. If a BLDC motor performance depends on the position of the rotor, based on the rotor location, the back EMF of the numerical expression can be obtained. When the BLDC rotates, the windings of the electromotive force or counter electromotive force is generated, which is opposed to the voltage supplied [25] to the winding of the main voltage. The polarity of the EMF is the opposite of the energized voltage. The modeling of the electromotive force assumes that the same back-EMF waveform is performed in all three stages. The rotor position based back EMF force and the expression of the BLDC motor is given below.

$$(emf)_a = f_a(\theta)K_{emf^{\omega}} \tag{9.1}$$

$$(emf)_b = f_b(\theta)K_{emf^{\omega}} \tag{9.2}$$

$$(emf)_c = f_c(\theta)K_{emf^{\omega}} \tag{9.3}$$

Among them, ω is the pointed velocity of the motor θ and K_{emf} is the propeller position of the EMF constant. Frame relays f_a (θ), f_b (θ), and f_c (θ) are functions of the rotor position. With different phase [26] motor voltages r, y, and b are described as follows:

$$\begin{bmatrix} V_a \\ V_b \\ V_c \end{bmatrix} = \begin{bmatrix} RA & 0 & 0 \\ 0 & RB & 0 \\ 0 & 0 & RC \end{bmatrix}\begin{bmatrix} I_a \\ I_b \\ I_c \end{bmatrix} + \begin{bmatrix} (L-m) & 0 & 0 \\ 0 & (L-m) & 0 \\ 0 & 0 & (L-m) \end{bmatrix}\begin{bmatrix} \frac{di_a}{dt} \\ \frac{di_b}{dt} \\ \frac{di_c}{dt} \end{bmatrix} + \begin{bmatrix} (emf)_a \\ (emf)_b \\ (emf)_c \end{bmatrix} \tag{9.4}$$

$$\frac{d_{\omega m}}{dt} = \left(\frac{P}{2_J}\right)(T_e - T_L - B_{\omega r}) \tag{9.5}$$

The torque based equation is given below:

$$T_e = K_{emf} (f_a (\theta) i_a + f_b (\theta) i_b + f_c (\theta) i_c) \quad (9.6)$$

The speed and torque relationships are given below:

$$\omega_m = \left(\frac{P}{2_J}\right)(T_e - T_L - B_{\omega r})dt \quad (9.7)$$

Equation (9.7) is used to analyze the BLDC motor speed in terms of load and torque.

9.3.2 Inverter Topology for BLDC Motor

Three-phase topologies are driven by excitation in two phases in time, producing the maximum amount of torque. Based on the rotor position feedback [27], the signal to the grid shown in Table 9.1 was created according to the fuzzy rule.

In addition, it must be considered that the freewheeling current flows through the diode as a result of phase inversion. A simplified schematic BLDC motor-based inverter system shown in Figure 9.3.

From Table 9.1 it is shown that at any given time the three phases of the two will be carried out in the third phase at the same time and will be closed [28] after the transition is settled. Ideally, a complete electrical cycle period at 360° and 120° can each show three-phase intervals and the interval in each half has a positive and a negative current.

The phase will remain closed outside this interval. In this 120-degree power mode, the BLDC motor model can be summarized as the based expression below:

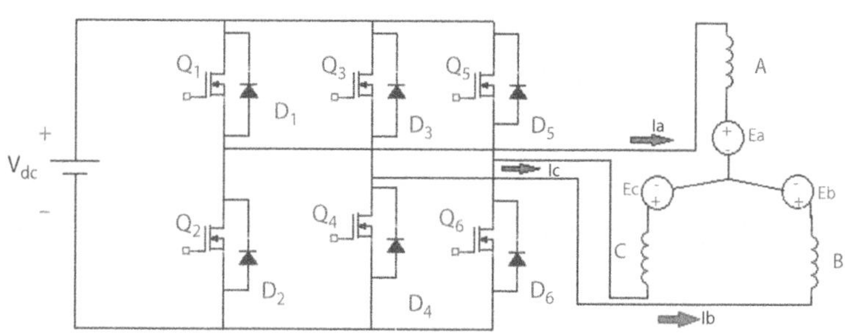

Figure 9.3 Inverter circuit diagram for BLDC Motor.

Table 9.1 AFOPC-based gate pulse sequence.

Switching interval	Sequence number	Switch closed		Phase current		
				A	B	C
0° – 60°	0	Q_1	Q_6	+1	Off	-1
60° – 120°	1	Q_1	Q_6	Off	+1	-1
120° – 180°	2	Q_1	Q_6	-1	+1	Off
180° – 240°	3	Q_1	Q_6	-1	Off	+1
240° – 300°	4	Q_1	Q_6	Off	-1	+1
300° – 360°	5	Q_1	Q_6	+1	-1	Off

$$\frac{di}{dt} = \frac{1}{L}\left(-R_p - K_{t\omega m} + V_{dc}\right) \tag{9.8}$$

$$\frac{d_{\omega m}}{dt} = \frac{1}{J}\left(K_{ti} - T_L - B_{\omega^m}\right) \tag{9.9}$$

where R = $2R_p$, L = $2L_S$, and i represents the equivalent DC current flowing through the on phase at any other time (other dimensions are defined as before).

9.3.2.1 *Adaptive Fuzzy Optimal Power Control (AFOPC) Based Speed Control of BLDCM*

This feature of the Adaptive Fuzzy Optimal Power Control (AFOPC) algorithm is simple and easy to comprehend. The design process of an Adaptive Fuzzy Optimal Power Control (AFOPC) process is in four stages [29] of operation: (i) Identifying input and output and the creation of discourse is determined, (ii) Fuzzification, (iii) Fuzzy rules definition, and (iv) Defuzzification.

(i) Identifying Input and Output and Creation of Discourse is Determined

The collection of all the information available on this concern is called the universal I/O values. The initial stage of the system in identifying [30] the

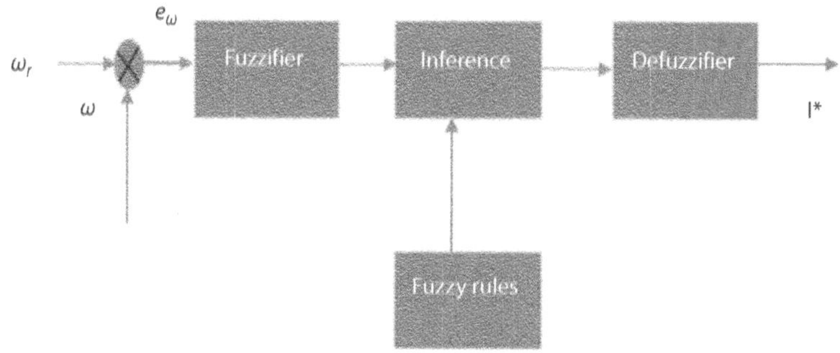

Figure 9.4 Adaptive fuzzy optimal power control (AFOPC) block.

inputs and outputs and specifying their range, thereby defining a position. The structure of the Adaptive Fuzzy Optimal Power Control (AFOPC) BLDC motor control block is represented in Figure 9.4.

In the primary function, the AFOPC regulator is defined to compute the speed error and e_ω is defined as $\omega_r - \omega$, where ω_r and ω represent the speed setpoint [31] and actual speed, respectively. The output current command I^* is used to deliver PWM to the system. Based on the motor's rating, the universal input values are selected at a lattice value of 1500 rpm.

(ii) Fuzzification

The method of converting instability into $\mu\ (e_\omega)$ fuzziness is called fuzzification. The membership degree of the quantified information into a fuzzy input set is defined in the membership function of a group. This degree of involvement is compared [32, 33] to the unit spacing [0, 1]. Then, fuzzy membership functions, adverse (A), Zero (Z), and positive (P) are defined and the process of the system is limited to three fuzzy variables. The output of the algorithm is the same as that defined within the system.

The below Figure 9.5 represents a high-speed error rate threshold value which is selected to ± 1RPM. The corresponding current threshold is selected as ±1A. Three-sided and trapezoidal membership functions are selected

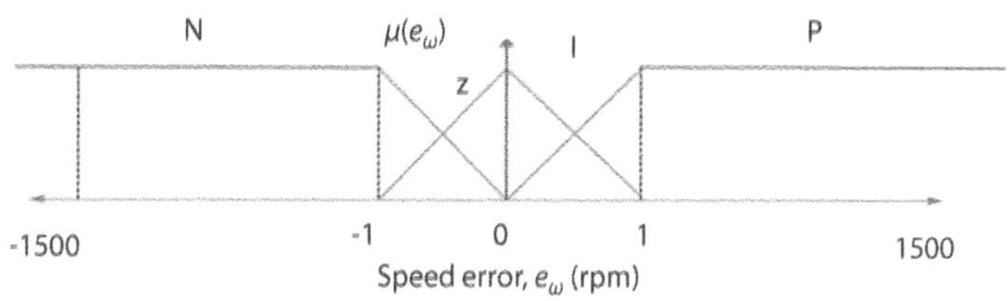

Figure 9.5 Input e_ω membership functions of BLDC motor.

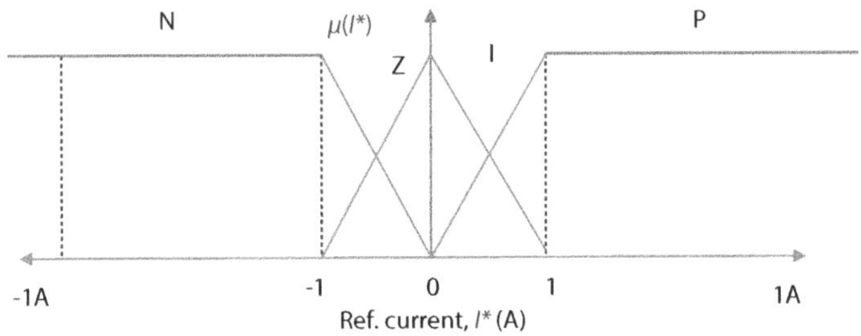

Figure 9.6 Output I^* membership function for BLDCM.

[34] for the two electronics. μ (e_ω) and μ (I^*) are depicted in Figures 9.5 and 9.6, where μ (e_ω) and μ (I^*) indicate that the membership degree of the velocity error e_ω and the reference current I^* are in their corresponding fuzzy sets.

(iii) Fuzzy Rules Definition

The definition of fuzzy rules is expressed by a group of "if-else" sentences of the system, establishing the relationship between input and output [35–48]. This means that even if not, then the form of rule-based deduction is known. These rules are critical to the vague causal process. In the next step, in the form of rule preparation, the "if-after" statement defines the specific control measure. A total of three rules were defined.

(iv) Defuzzification

Since the input and the output of the actual system are unambiguous values, the fuzzy process requires the output to be converted [44] from a fuzzy rule system. The method used for defuzzification can be selected from various methods [45] including the principle of maximum membership, the method of the center of gravity, weighted average, and maximum average principle. The membership function [49, 50] for each fuzzy input-output must correspond to the rules to be combined. Based on the selected rule, a degree corresponding to the number of members of each of the fuzzy mapping input directly to the fuzzy set's output is called a recommendation by the subsequent [51–57] process, finally resulting in a fuzzy set. The methods of truncation and centroid have meaning and fuzzification, respectively.

The AFOPC controller input below has no-load and weight conditions for output calculation processing. The third fuzzy input, shown in Figures 9.7(a)&(b), is set to a Negative (N) and it is is not revealed because its membership function value is zero and there is no output to the fuzzy set.

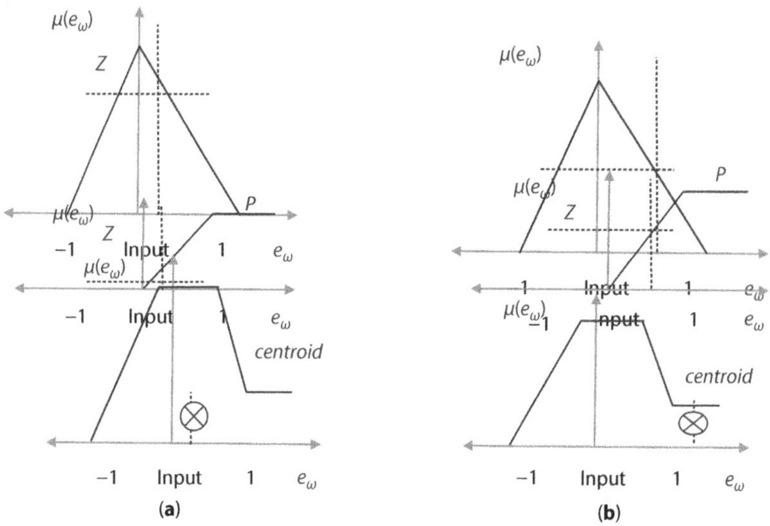

Figure 9.7 (a) No load; (b) full load.

9.3.2.2 *Adaptive Fuzzy Optimal Power Control (AFOPC)*

Step 1: Set the error to zero and the weight W. A) Error, b) Medium error rate (change in error) and weight.
Step 2: Read the fuzzy membership functions defined in the process of a three fuzzy variable system that is defined in the Negative (N), Zero (Z), and Positive (P).
Step 3:

$I^* \cong 2I_r$	If $e_\omega \geq 1\ rpm$
$I^* \cong 2I_r$	If $e_\omega \leq -1\ rpm$
$-2I_r \leq I^* \leq 2I_r$	If $0 \leq \lvert e_\omega \rvert \leq 1\ rpm$

Step 4: Compute the speed error e_ω, defined as $\omega_r - \omega$
Step 5: Evaluate the control law $\mu(e_\omega)$
Step 6: Calculate the weight update, that is vector $\dot{W}(k) = \Delta\dot{W}$.
Step 7: Update the vector of weights used, $\dot{W}(k) = \dot{W}(k-1) + \Delta\dot{W}$.
Step 8: This vector of weights membership is mapped to the unit interval [0,1]

Case 1: If Error_in_Speed is NB == 1 and Change_in_error is N, Z, B, then at that point position is Negative (N), Zero (Z), and Positive (P).

Case 2: If Error_in_Speed is NM == 1 and Change_in_error is N, Z, B, then at that point position is Negative (N), Zero (Z), and Positive (P) until the stop request is received. The best performance can be achieved until satisfaction.

Step 9: Centroid calculation is for each output set. The centroid fuzzy set is Y=0.

Step 10: Defuzzification

$$y = \frac{\sum\limits_{i=1,3} | y = y^{(i)} | * y^{(i)}}{\sum\limits_{i=1,3} | y = y^{(i)} |} \tag{9.10}$$

This AFOPC control scheme may be used alone in the application and it is acceptable. It requires less computational complexity and low steady-state error.

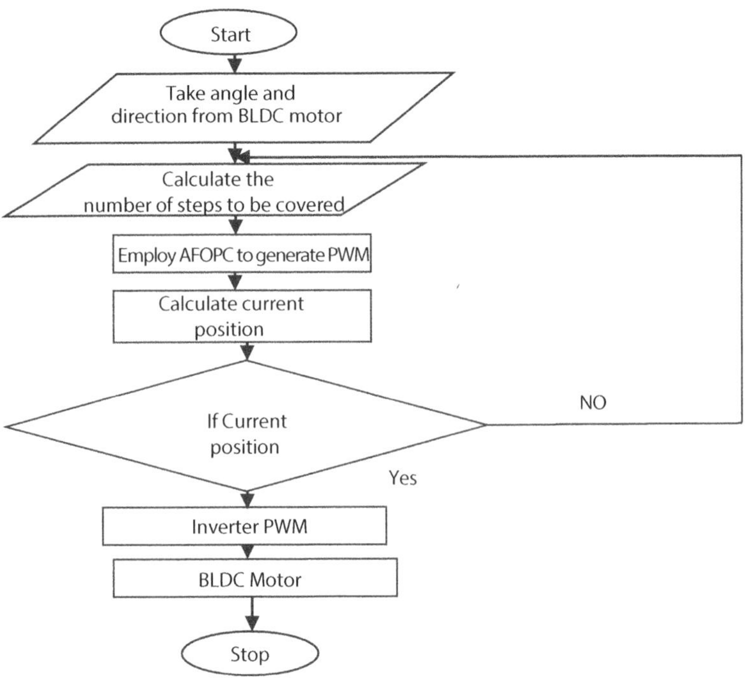

Figure 9.8 Flow chart for proposed AFOPFC technique.

Above, Figure 9.8 describes the flow chart for the proposed Adaptive Fuzzy Optimal Power Control (AFOPC) is clearly depicted.

9.4 Result and Argument

The recreation is carried out with the help of the MATLAB/SIMULINK program. Simulink library files, simulation model files for Adaptive Fuzzy Optimal Power Control (AFOPC), and simulation model files for Adaptive Fuzzy Optimal Power Control (AFOPC) including inbuilt models, as seen in Figure 9.9. In this experiment, we investigate how well Adaptive Fuzzy Optimal Power Management (AFOPC) can control the speed of a Brushless DC motor under different loads and torques.

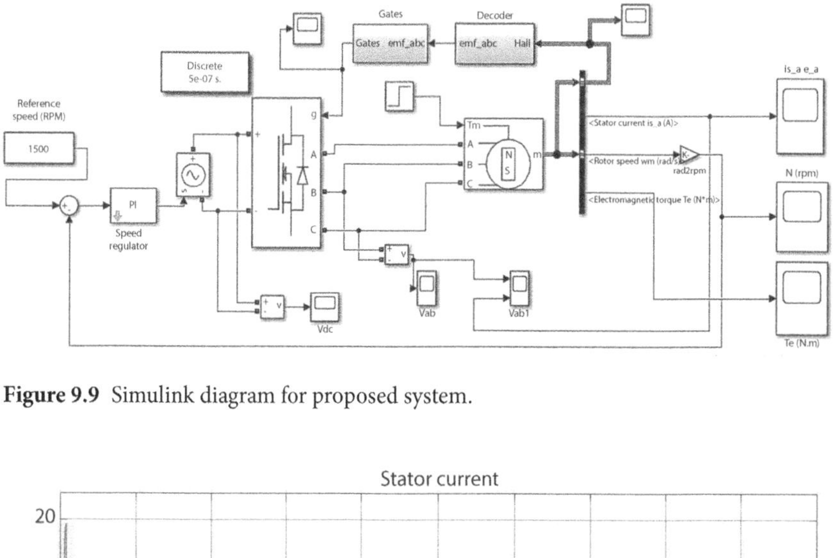

Figure 9.9 Simulink diagram for proposed system.

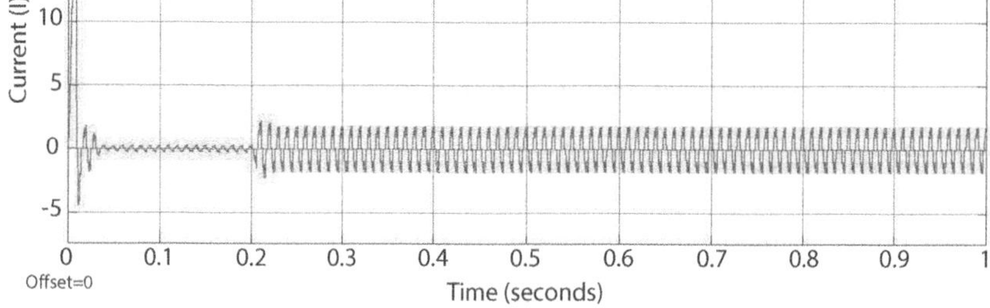

Figure 9.10 Output current of projected system.

Figure 9.9 represents the Simulink model for the BLDC motor control using an Adaptive Fuzzy Optimal Power Control (AFOPC). Figure 9.10 describes the output current of projected system with the help of the MATLAB/SIMULINK program. Output of the actual system are unambiguous values, the fuzzy process requires the output to be converted from a fuzzy rule system.

9.5 Conclusions

The proposed Adaptive Fuzzy Optimal Power Controller (AFOPC) is designed to achieve the lowest possible losses while maintaining the highest possible motor speed. Taking into consideration the ambiguous regulations, the link between static currents is discovered in order to reduce power loss and acquire orientation in the field by maintaining it in a sober condition. The success of a nonlinear controller is completely independent of the variables affecting the system. Changes in current and speed always result in the formation of a regulated duty rotation signal within the range before the linear modulation region is determined. The stability system in the non-linear closed spiral system is proven using the whole system dynamics and controller design. The state is surrounded by the desired equilibrium and is guaranteed to be converged in this system. Adding speed stability analysis, which is valid, the AFOPC controller can be readily performed with the system's speed response and the AFOPC controller scheme gives superior results with a steady-state error of 0.34 percent, which is lower than the industry standard.

References

1. Trivedi, M. S., & Keshri, R. K. (2020) "Evaluation of Predictive Current Control Techniques for PM BLDC Motor in Stationary Plane" IEEE Access, Vol.8, PP.46217–46228. doi:10.1109/access.2020.2978695.
2. Valle, R. L., De Almeida, P. M., Fogli, G. A., Ferreira, A. A., & Barbosa, P. G. (2020)"Simple and Effective Digital Control of a Variable-Speed Low Inductance BLDC Motor Drive" IEEE Access, Vol.8,PP.13240–13250. doi:10.1109/access.2020.2966437.
3. Chen, X., & Liu, G. (2019) "Sensorless Optimal Commutation Steady Speed Control Method for a non-ideal Back-EMF BLDC Motor Drive System including Buck Converter" IEEE Transactions on Industrial Electronics, Vol.67,No.7,PP.6147 – 6157, doi:10.1109/tie.2019.2945282.

4. Macahig, N. A. (2020)"A 6-Wire 3-Phase Inverter Topology for Improved BLDC Performance and Harmonics" 2020 IEEE Applied Power Electronics Conference and Exposition (APEC). doi:10.1109/apec39645.2020.9124358.
5. Mahmud, M., Motakabber, S. M. A., ZahirulAlam, A. H. M., &Nordin, A. N. (2020) "Adaptive PID Controller Using for Speed Control of the BLDC Motor" 2020 IEEE International Conference on Semiconductor Electronics (ICSE). doi:10.1109/icse49846.2020.9166883.
6. Devarapalli, R., Naga Lakshmi, N. J., & Prasad, U. (2020)"Application of a Novel Political Optimization in Optimal Parameter Design of PI Controller for the BLDC motor Speed Control" 2020 International Conference on Emerging Frontiers in Electrical and Electronic Technologies (ICEFEET). doi:10.1109/icefeet49149.2020.9186957.
7. Tian, B., An, Q.-T., &Molinas, M. (2019)"High-Frequency Injection-Based Sensorless Control for a General Five-Phase BLDC Motor Incorporating System Delay and Phase Resistance" IEEE Access, Vol.7, PP.162862–162873. doi:10.1109/access.2019.2950256.
8. Lee, W., Kim, J. H., Choi, W., &Sarlioglu, B. (2018) "Torque Ripple Minimization Control Technique of High-Speed Single-Phase Brushless DC Motor Drive for Electric Turbocharger" IEEE Transactions on Vehicular Technology, Vol.67, No.11, PP.10357 – 10365,doi: 10.1109/TVT.2018.2866779.
9. Chen, S., Liu, G., & Zhu, L. (2017)"Sensorless Control Strategy of a 315 kW High-Speed BLDC Motor Based on a Speed-Independent Flux Linkage Function" IEEE Transactions on Industrial Electronics, Vol.64, No.11, PP.8607–8617.
10. Li, W., Fang, J., Li, H., & Tang, J. (2016)"Position Sensorless Control without Phase Shifter for High-Speed BLDC Motors with Low Inductance and Nonideal Back EMF" IEEE Transactions on Power Electronics, Vol.31, No.2, PP.1354–1366.
11. Baszynski, M., &Pirog, S. (2014) "A Novel Speed Measurement Method for a High-Speed BLDC Motor Based on the Signals from the Rotor Position Sensor" IEEE Transactions on Industrial Informatics, Vol.10, No.1, PP.84–91. doi:10.1109/tii.2013.2243740.
12. Ozturk, S. B., &Toliyat, H. A. (2011)"Direct Torque and Indirect Flux Control of Brushless DC Motor" IEEE/ASME Transactions on Mechatronics, Vol.16, No.2, PP.351–360,doi: 10.1109/TMECH.2010.2043742.
13. Rodriguez, F., &Emadi, A. (2007)"A Novel Digital Control Technique for Brushless DC Motor Drives" IEEE Transactions on Industrial Electronics, Vol.54, No.5, PP. 2365–2373.
14. Stirban, A., Boldea, I., &Andreescu, G.-D. (2012) "Motion-Sensorless Control of BLDC-PM Motor with Offline FEM-Information-Assisted Position and Speed Observer" IEEE Transactions on Industry Applications, Vol.48,No.6,PP.1950–1958,doi: 10.1109/TIA.2012.2226194.

15. Mishra, P., Banerjee, A., &Ghosh, M. (2020) "FPGA Based Real-Time Implementation of Quadral-Duty Digital PWM Controlled Permanent Magnet BLDC Drive" IEEE/ASME Transactions on Mechatronics, Vol.25, No.3, PP.1456 – 1467,doi:10.1109/tmech.2020.2977859.
16. Adil Usman., Bharat Singh Rajpurohit (2019) "Comprehensive Analysis of Demagnetization Faults in BLDC Motors Using Novel Hybrid Electrical Equivalent Circuit and Numerical Based Approach" IEEE Access,Vol.7,PP.147542 - 147552,doi: 10.1109/ACCESS.2019.2946694.
17. M, P. M., & S, A. E. xavier. (2018) "Design of speed control and reduction of torque ripple factor in BLDC motor using spider-based controller" IEEE Transactions on Power Electronics, Vol.34, No.8,PP.7826 - 7837 doi:10.1109/tpel.2018.2880916.
18. Zhou, X., Chen, X., Peng, C., & Zhou, Y. (2018) "High Performance Non-Salient Sensorless BLDC Motor Control Strategy from Standstill to High Speed" IEEE Transactions on Industrial Informatics, Vol.14, No.10, PP. 4365 – 4375,doi: 10.1109/TII.2018.2794461.
19. Feng, J., Liu, K., & Wang, Q. (2018)"Scheme based on buck-converter with three-phase H-bridge combinations for high-speed BLDC motors in aerospace applications" IET Electric Power Applications, Vol.12, No.3, PP.405–414. doi:10.1049/iet-epa.2017.0615.
20. Nerat, M., &Vrancic, D. (2016) "A Novel Fast-Filtering Method for Rotational Speed of the BLDC Motor Drive Applied to Valve Actuator" IEEE/ASME Transactions on Mechatronics, Vol.21, No.3, PP.1479–1486,doi: 10.1109/TMECH.2015.2505321.
21. Bist, V., & Singh, B. (2014)"An Adjustable-Speed PFC Bridgeless Buck-Boost Converter-Fed BLDC Motor Drive" IEEE Transactions on Industrial Electronics, Vol.61, No.6, PP. 2665–2677. doi:10.1109/tie.2013.2274424.
22. Jayachandran, A., and D. Stalin David. "Textures and Intensity Histogram Based Retinal Image Classification System Using Hybrid Colour Structure Descriptor." Biomedical and Pharmacology Journal, vol. 11, no. 1, 2018, p. 577+. Accessed 12 Feb. 2021.
23. D. Stalin David, 2019, "Parasagittal Meningiomia Brain Tumor Classification System based on MRI Images and Multi Phase level set Formulation", Biomedical and Pharmacology Journal, Vol.12, issue 2, pp.939-946.
24. D. S. David and A. Jeyachandran, "A comprehensive survey of security mechanisms in healthcare applications," 2016 International Conference on Communication and Electronics Systems (ICCES), Coimbatore, 2016, pp. 1-6, doi: 10.1109/CESYS.2016.7889823.
25. Stalin David, D., Jayachandran, A. A new expert system based on hybrid colour and structure descriptor and machine learning algorithms for early glaucoma diagnosis. Multimed Tools Appl 79, 5213–5224 (2020). https://doi.org/10.1007/s11042-018-6265-1.

26. D Stalin David, A Jayachandran, 2018, Robust Classification of Brain Tumor in MRI Images using Salient Structure Descriptor and RBF Kernel-SVM, TAGA Journal of Graphic Technology, Volume 14, Issue 64, pp.718-737.
27. D Stalin David, 2016, Robust Middleware based Framework for the Classification of Cardiac Arrhythmia Diseases by Analyzing Big Data, International Journal on Recent Researches In Science, Engineering & Technology, 2018, Volume 4, Issue 9, pp.118-127.
28. M. Rajdhev, D. Stalin David, "Internet of Things for Health Care", International Journal of Scientific Research in Computer Science, Engineering and Information Technology (IJSRCSEIT), ISSN : 2456-3307, Volume 2 Issue 2, pp. 800-805, March-April 2017.
29. P. Prasanth, D. Stalin David, "Defensing Online Key detection using Tick Points", International Journal of Scientific Research in Computer Science, Engineering and Information Technology (IJSRCSEIT), ISSN : 2456-3307, Volume 2 Issue 2, pp. 758-765, March-April 2017.
30. A. Sudalaimani, D. Stalin David, "Efficient Multicast Delivery for Data Redundancy Minimization over Wireless Data Centres", International Journal of Scientific Research in Computer Science, Engineering and Information Technology (IJSRCSEIT), ISSN : 2456-3307, Volume 2 Issue 2, pp. 751-757, March-April 2017.
31. R. Abish, D. Stalin David, "Detecting Packet Drop Attacks in Wireless Sensor Networks using Bloom Filter", International Journal of Scientific Research in Computer Science, Engineering and Information Technology (IJSRCSEIT), ISSN : 2456-3307, Volume 2 Issue 2, pp. 730-735, March-April 2017.
32. A. Vignesh, D. Stalin David, "Novel based Intelligent Parking System", International Journal of Scientific Research in Computer Science, Engineering and Information Technology (IJSRCSEIT), ISSN : 2456-3307, Volume 2 Issue 2, pp. 724-729, March-April 2017.
33. D Stalin David, 2020, 'Diagnosis of Alzheimer's Disease Using Principal Component Analysis and Support Vector Machine, International Journal of Pharmaceutical Research, Volume 12, Issue 2, PP.713-724.
34. Jaswanth K S, Dr. D. Stalin David, "A Novel Based 3d Facial Expression Detection Using Recurrent Neural Network", International Journal of Scientific Research in Computer Science, Engineering and Information Technology (IJSRCSEIT), ISSN : 2456-3307, Volume 6 Issue 2, pp. 48-53, March-April 2020.
35. D Stalin David, 2020, 'An Intellectual Individual Performance Abnormality Discovery System in Civic Surroundings' International Journal of Innovative Technology and Exploring Engineering, Volume 9, Issue 5, PP.2196-2206.
36. D Stalin David, 2020, 'Machine learning for the prelude diagnosis of dementia', International Journal of Pharmaceutical Research, Volume 13, Issue 3, PP.2329-2335.

37. David, D.S. and Y. Justin, 2020. A Comprehensive Review on Partition of the Blood Vessel and Optic Disc in Retinal Images. Artech J. Eff. Res. Eng. Technol., 1: 110-117.
38. D. Stalin David and A.A. Jose, 2020. Retinal image classification system for diagnosis of diabetic retinopathy using SDC Methods. Artech J. Eff. Res. Eng. Technol., 1: 87-93.
39. D. Stalin David and T. Joseph George, 2020. Identity-based Sybil attack detection and localization. Artech J. Eff. Res. Eng. Technol., 1: 94-98.
40. David, D.S. and L. Arun, 2020. Classification of brain tumor type and grade using MRI texture and shape in a machine learning scheme. Artech J. Eff. Res. Eng. Technol., 1: 57-63.
41. David, D.S., 2020. Retinal image classification system for diagnosis of diabetic retinopathy using morphological edge detection and feature extraction techniques. Artech J. Eff. Res.Eng. Technol., 1: 28-33.
42. David, D.S., 2020. A novel specialist system based on hybrid colour and structure descriptor and machine learning algorithms for early diabetic retinopathy diagnosis. Artech J. Eff. Res. Eng. Technol., 1: 50-56.
43. David, D.S. and M. Samraj, 2020. A comprehensive survey of emotion recognition system in facial expression. Artech J. Eff. Res. Eng. Technol., 1: 76-81.
44. David, D.S. and L. Arun, 2020. Multi-view 3D face renovation with deep recurrent neural networks. Artech J. Eff. Res. Eng. Technol., 1: 64-68.
45. David, D.S. and S. Namboodiri, 2020. Improvement of framework for the grouping of CA diseases by investigating big data. Artech J. Eng. Appl. Technol., 1: 7-14.
46. Stalin David D , Saravanan M, 2020, 'Multi-perspective DOS Attack Detection Framework for Reliable Data Transmission in Wireless Sensor Networks based on Trust', International Journal of Future Generation Communication and Networking , Volume 13, Issue 4, PP.1522–1539.
47. J. K. S and D. S. David, "A Novel Based 3D Facial Expression Detection Using Recurrent Neural Network," 2020 International Conference on System, Computation, Automation and Networking (ICSCAN), Pondicherry, India, 2020, pp. 1-6, doi: 10.1109/ICSCAN49426.2020.9262287.
48. Stalin David D, Saravanan M, "Enhanced Glaucoma Detection Using Ensemble based CNN and Spatially Based Ellipse Fitting Curve Model", Solid State Technology, Volume 63, Issue 6, PP.3581-3598.
49. Stalin David D, Saravanan M, Jayachandran A, "Deep Convolutional Neural Network based Early Diagnosis of multi class brain tumour classification", Solid State Technology, Volume 63, Issue 6, PP.3599-3623.
50. D. Jayakumar; Dr. U. Palani; D. Raghuraman; Dr. D. Stalin David; D. Saravanan; R. Parthiban; S. Usharani. "Certain Investigation On Monitoring The Load of Short Distance Orienteering Sports On Campus Based On Embedded System Acceleration Sensor". European Journal Of Molecular & Clinical Medicine, 7, 9, 2021, 2477-2494.

51. R. Parthiban; S. Usharani; D. Saravanan; D. Jayakumar; Dr. U. Palani; Dr. D. Stalindavid; D. Raghuraman. "Prognosis Of Chronic Kidney Disease (Ckd) Using Hybrid Filter Wrapper Embedded Feature Selection Method". European Journal Of Molecular & Clinical Medicine, 7, 9, 2021, 2511-2530.
52. Dr. U. Palani; D. Raghuraman; Dr. D. Stalindavid; R. Parthiban; S. Usharani; D. Jayakumar; D. Saravanan. "An Energy-Efficient Trust Based Secure Data Scheme In Wireless Sensor Networks". European Journal Of Molecular & Clinical Medicine, 7, 9, 2021, 2495-2510.
53. Dr. D. Stalin David; R. Parthiban; D. Jayakumar; S. Usharani; D. Raghu Raman; D. Saravanan; Dr.U. Palani. "Medical Wireless Sensor Network Coverage And Clinical Application Of MRI Liver Disease Diagnosis". European Journal Of Molecular & Clinical Medicine, 7, 9, 2021, 2559-2571.
54. D. Raghu Raman; D. Saravanan; R. Parthiban; Dr. U. Palani; Dr. D. Stalin David; S. Usharani; D. Jayakumar. "A Study On Application Of Various Artificial Intelligence Techniques On Internet Of Things". European Journal Of Molecular & Clinical Medicine, 7, 9, 2021, 2531-2557.
55. D. Saravanan; Dr. D. Stalin David; S. Usharani; D. Raghuraman; D. Jayakumar; Dr. U. Palani; R. Parthiban. "An Energyefficient Traffic-Less Channel Scheduling Based Data Transmission Inwireless Networks". European Journal Of Molecular & Clinical Medicine, 2020, Volume 7, Issue 11, Pages 5704-5722.
56. S. Usharani; D. Jayakumar; Dr. U. Palani; D. Raghuraman; R. Parthiban; D. Saravanan; Dr. D. Stalin David. "Industrialized Service Innovation Platform Based On 5g Network And Machine Learning". European Journal Of Molecular & Clinical Medicine, 2020, Volume 7, Issue 11, Pages 5684-5703.
57. Dr. D. Stalin David, Mr. D. Saravanan, "Certain Investigation On Iot Therapeutic Image Recognition and Rivaroxabanpreclude Thrombosis In Patients", 2021, Pg.No:51-66, ISBN: 978-81-948555-1- 4

10

Fuzzy Fractional Order PID Controller Design for Single Link Robotic Arm Manipulator

Sayan Das[1] and Naiwrita Dey[2]*

[1]Department of AEIE, RCCIIT, Kolkata, India
[2]Department of ECE, RCCIIT, Kolkata, India

Abstract
This chapter is about fuzzy gain scheduling and order optimization of a proposed Fractional Order PID (FOPID) model. A simple Mamdani type Fuzzy Inference System (FIS) is developed with five different rule bases to select values of three PID gains and orders of integrator and differentiator depending upon error and rate of change in error. The design of the controller mentioned above is discussed elaborately. A comparative survey between a classical and Fuzzy FOPID controller has been observed and analyzed with a single-link robotic arm manipulator system. Performance analysis of the proposed controller is analyzed in the time domain. Different performance indices are also calculated. Step response shows the enhanced performance of the proposed Fuzzy Fractional-order PID (FFOPID) controller over the conventional PID and FOPID controller. The stability condition is obtained for the proposed controller. The simulation study is carried out using MATLAB.

Keywords: Fractional calculus, Fuzzy, PID, MATLAB, robotic arm manipulator

10.1 Introduction

In recent years, the importance of robotics in the industrial field is growing immensely. Sophisticated robotic arms are now being used in different industrial applications like painting, welding, filtering out worst

**Corresponding author*: naiwritadey@gmail.com

Vipin Chandra Pal, Suman Lata Tripathi, and Souvik Ganguli (eds.) Industrial Control Systems, (203–230) © 2024 Scrivener Publishing LLC

products quickly, and performing different heavy and dangerous jobs which are quite impossible to perform using manpower. Not only in industry, robotic arms are extensively used in different space programs, nuclear reactors, deep ocean surveys, etc. For performing such jobs, robotic arms need to act accurately and instantaneously, otherwise a very small error can lead to the worst consequences. These agendas have opened a research paradigm for robot arm control [1]. Robust control of flexible link manipulators has become a blooming field of research in robotic control system development. Robot arms are highly nonlinear and unstable systems. Controlling such a system is challenging for any controller. In literature, different control techniques are discussed for robotic arm manipulators. Fuzzy logic based controllers have gained massive popularity in the robotics industry due to its efficiency and simple design structure. Different fuzzy models such as the Mamdani model or Takagi-Sugeno model based applications have been reported in the literature [2]. Also, fuzzy inference systems can be designed accordingly to attain the control objectives of stability and robustness [3, 4]. PID controllers are the most common for feedback control. Since its introduction, classical PID controllers have been developed by means of more stability, accuracy, and robustness [5, 6]. For improving system behavior, a generalized version of a conventional PID controller has been developed by implementing non-integer orders of integration and differentiation, which is known as a Fractional-order PID (FOPID) controller [7]. Fractional-order PID controllers offer more accuracy in controlling [8, 9]. Hence, FOPID controllers give more adaptable time domain and frequency domain responses to the control system with a fulfillment of better and robust performance [10]. In recent research works, a more general FOPID controller has been developed by introducing a nonlinear term which varies depending upon the error, controlling the rate of convergence as per need [11, 12]. Nonlinear FOPID controllers show better performance in controlling highly nonlinear robotic systems with higher degrees of freedom [13]. After implementation of fractional order calculus in control systems, several fractional order calculus based control techniques have been introduced. FOPID control is one of them. The literature shows FOPID controllers are very efficient in controlling robot arms. This also makes the manipulator system capable of disturbance-rejection to some extent. In this chapter, the disadvantages of implementing FOPID controllers in manipulator systems are taken into account from the literature and accordingly, a more adaptable Fuzzy FOPID controller is designed mainly for manipulator position control.

Also, the proposed FFOPID controller is tested on a single-link rigid manipulator model to review its performance.

This chapter is focuses on developing a fuzzy logic based FOPID controller mainly for robot manipulator applications. The design of the controller mentioned above is discussed elaborately. A comparative survey between classical Fuzzy and Fuzzy FOPID controllers has been observed and analyzed with a single-link robotic arm manipulator system. Rise time, settling time, and percentage of peak overshoot is obtained to evaluate the time domain performance of the proposed controller. Performance indices such as ISE, IAE, and ITAE are also calculated. Step response shows the improved performance of the Fuzzy FOPID controller over the conventional Fuzzy PID controller. Stability condition is also obtained for the proposed controller.

10.2 Fuzzy Logic Control

The concept of a fuzzy set was first proposed by L. Zadeh in 1965. It is very similar to human reasoning and can be applied to solve logical problems with uncertain information as it considers absolute true or absolute false along with the intermediate values [14]. Mamdani type Fuzzy Systems and Takagi-Sugeno type Fuzzy Systems are the two most popular type of classifications of fuzzy system. Basic structure of Mamdani type FIS is discussed here [15].

10.2.1 Mamdani Type Fuzzy System

The Mamdani type fuzzy system is a nonlinear mapping from $U = U_1 \times \cdots \times U_L$, where $u_1 \in U_1, \cdots, u_L \in U_L$ and $y \in Y$ to Y. The general rule-form in case of the Mamdani systems is:

$$R^i : \text{IF } \bar{u}_1 \text{ is } \bar{A}_1^j \text{ and } \bar{u}_2 \text{ is } \bar{A}_2^k \text{ and} \ldots \text{and } \bar{u}_L \text{ is } \bar{A}_L^l, \text{ THEN } \bar{y} \text{ is } \bar{C}_i, \tag{10.1}$$

where $i \in [1, N], A_1^j$ represents the linguistic value of the j th term in u_1 and so on, and L represents the number of premise variables.

Singleton fuzzification and minimum as "and" operation is used in this system. The confidence of each rule for this system is obtained from:

$$\lambda_i(u) = \min\left(\mu_{A_1^j}(u_1), \cdots, \mu_{A_L^l}(u_L)\right) \tag{10.2}$$

The obtained output fuzzy set by using the Mamdani implication is:

$$\mu_{C_i'}(y) = \min\left(\lambda_i(u), \mu_{C_i}(y)\right) \tag{10.3}$$

Now, considering the logical relations among the rules as "or" and using the maximum for that, the overall Mamdani output fuzzy set is determined:

$$C' = \bigcup_{i=1}^{N} C_i' \tag{10.4}$$

The membership function for the same is calculated by:

$$\mu_{C'}(y) = \max_{i=1}^{N}\left(\mu_{C_i'}(y)\right) \tag{10.5}$$

At last, the crisp output is obtained using the Center-of-Gravity defuzzification approach.

10.3 Fractional Order Proportional Integral Derivative (FOPID) Controller

Fractional Order PID controllers are an extension of Conventional PID controllers [16]. FOPID is developed by implementing fractional order calculus in the conventional PID controller equation. FOPID controllers are more robust due to their higher degree of freedom [17]. In this section, the basic theory behind FOPID controller design and its different types is discussed.

10.3.1 Introduction to Fractional-Order Calculus

Fractional order calculus is basically the generalization of ordinary integration and differentiation, where order of the derivative can be an integer, fraction, or complex number [18–20]. In the following sub-section, a brief overview of different fractional order mathematical operators are discussed.

10.3.1.1 Fractional-Order Differintegral Operator

A Fractional Differintegral operator is the combination of differentiator and integrator, D_t^α of order $\alpha \in \mathbb{R}$. For $\alpha < 0$, it acts as an integrator and for $\alpha > 0$, it acts as a differentiator. This operator is defined as:

$$D_t^\alpha f(t) = \begin{cases} \int_0^t f(\tau) d\tau^\alpha & \alpha < 0, \\ f(t) & \alpha = 0, \\ \dfrac{d^\alpha f(t)}{dt^\alpha} & \alpha > 0, \end{cases} \tag{10.6}$$

The fractional order differintegrator in Equation 10.6 can be mathematically formulated in many different ways. Grünwald-Letnikov as GL and Riemann-Liouville as RL definitions are described below:

Riemann-Liouville Definition: The left fractional order derivative according to Riemann-Liouville's definition.

$$D_t^\alpha f(t) = \frac{1}{\Gamma(n-\alpha)} \frac{d^n}{dt^n} \int_0^t f(\tau)(t-\tau)^{n-\alpha-1} d\tau, \qquad \alpha \in (n-1, n], n \in \mathbb{Z}^+ \tag{10.7}$$

Here, Γ (.) is the Gamma function.

Grünwald-Letnikov Definition:

$$D_t^\alpha f(t) \approx \frac{1}{h^\alpha} \sum_{j=0}^{t/h} \omega_j^{(\alpha)} f(t-jh), \qquad \alpha \in \mathbb{R} \tag{10.8}$$

Here, h represents the step-size and ω_j represents some coefficients calculated as:

$$\omega_0^{(\alpha)} = 1; \quad \omega_j^{(\alpha)} = \left(1 - \frac{\alpha+1}{j}\right) \omega_{j-1}^{(\alpha)}, j = 1,2,3,4,... \tag{10.9}$$

The RL definition of a fractional-order derivative lacks physical applications due to its need for initial conditions. Caputo modified RL definition Equation 10.7 later. From the definitions, it is clear that the GL definition is the simplest, most popular, and is the most easily implementable as an algorithm. Throughout this chapter, the Grünwald-Letnikov definition of the fractional differintegrator operator is referred to.

10.3.1.2 Laplace Transform of Fractional Differintegrator

The Laplace transform of fractional differintegrator in Equation 10.8 with zero initial condition is expressed as:

$$\mathcal{L}\{D_t^{\alpha} f(t)\} = s^{\alpha} F(s) - \sum_{k=0}^{\alpha-1} s^{\alpha-k-1} f^{(k)}(0)$$

For zero initial condition,

$$\mathcal{L}\{D_t^{\alpha} f(t)\} = s^{\alpha} F(s), \qquad \alpha \in \mathbb{R} \tag{10.10}$$

Here, $F(s)$ represents the Laplace transform of $f(t)$.

10.3.1.3 Approximation Methods of Fractional-Order Laplace Transform

Implementation of fractional order control applications is involved in various approximation techniques. Use of the Oustaloup approximation, refined Oustaloup approximation, Matsuda approximation, the power series expansion technique, transfer function estimation, curve fitting technique, etc. has been found in the literature [21, 22]. The following sub-section has reported two very popular approximation techniques.

Oustaloup Approximation

This is an approximation technique for approximating s^{α} within the preferred range of frequency scale (ω_L, ω_H). It is given below:

$$s^{\alpha} \approx \omega_h^{\alpha} \prod_{k=1}^{N} \frac{s+\omega_k^z}{s+\omega_k^p}, \quad 0<\alpha<1 \tag{10.11}$$

Here, N means the order of approximation.
ω_k^z and ω_k^p are the zeros and poles respectively. They are computed as:

$$\omega_k^z = \omega_L \left(\frac{\omega_H}{\omega_L} \right)^{\frac{2k-1-\alpha}{2N}} \tag{10.12}$$

$$\omega_k^p = \omega_L \left(\frac{\omega_H}{\omega_L} \right)^{\frac{2k-1+\alpha}{2N}} \tag{10.13}$$

Refined Oustaloup Approximation:
It is the refined version of the Oustaloup approximation. It is more robust. It is defined as:

$$s^\alpha \approx \left(\frac{a\omega_h}{b} \right)^\alpha \left(\frac{as^2 + b\omega_h s}{a(1-\alpha)s^2 + b\omega_h s + a\alpha} \right) \prod_{k=-N}^{N} \frac{s+\omega_k^z}{s+\omega_k^p}, \quad 0<\alpha<1 \tag{10.14}$$

Approximation constants a and b are selected as 9 and 10, respectively, for achieving better approximation. Present work has approximated the fractional order derivatives and integrals by means of Refined Oustaloup Approximation.

10.3.2 Fractional-Order PID Controller-FOPID

The conventional PID controller can be expressed with the equation below:

$$C(s)_{PID} = \left(K_P + \frac{K_I}{s} + K_D s \right) \cdot E(s) \tag{10.15}$$

Here, $C(s)_{PID}$ is the controller output and $E(s)$ is the error in the plant output.

K_P(Proportional), K_I (Integral), and K_D (Derivative) are the PID controller gain parameters. By adjusting these gains the controller action can be controlled. The three adjustable gains associated with these terms depend on the error magnitude, time needed to nullify accumulated error, and error prediction horizon. Conventional PID controllers have three degrees of freedom as we can manipulate the controller action via changing these three gain constants.

Fractional calculus is used to extend further the design of the conventional PID controller [23]. The integral and derivative control actions can be the order of any real number. Usually, it is chosen between 0 and 2. It gives more degrees of freedom and robustness for FOPID controllers because of its non-integer order and the integral term acts with selective part to the accumulated error and based on the differentiation order value, hence different prediction horizons [24]. Predicted values of the error are referred by derivative control action. There are many forms of FOPID controllers available out there [25, 26].

10.3.2.1 Podlubny's FOPID ($PI^{\lambda} D^{\mu}$) Controller

It is the most basic form of FOPID controller proposed by Podlubny. It is obtained by ordering the derivative and integral action with μ and λ respectively. The controller equation for the same is shown below:

$$C(s)_{FOPID} = \left(K_P + \frac{K_I}{s^{\lambda}} + K_D s^{\mu} \right) \cdot E(s), \quad \lambda, \mu \in [0,2] \qquad (10.16)$$

M. Tenoutit's FO-$(PID)^N$ Controller: This form of FOPID was proposed by M. Tenoutit. This form is expressed as:

$$C(s)_{(PID)^N} = \frac{1}{s^N} \left(K_P + \frac{K_I}{s} + K_D s \right) \cdot E(s), \quad N \in [0,1] \qquad (10.17)$$

Here, N is the order.

10.3.2.2 Internal Mode Control (IMC) Based FOPID Controller

Tavakoli proposed this controller, given by the following expression:

$$C(s)_{IMCFOPID} = \left(K_P + \frac{K_I}{s^{\alpha}} + \frac{K_D s^{\alpha}}{\gamma s^{\alpha} + 1} \right) \cdot E(s), \quad \alpha \in [0,1] \qquad (10.18)$$

Here, α is the order. The denominator in the derivative part of the equation is appended to make it a proper transfer function.

Table 10.1 Types of FOPID controllers and their structures.

Type of FOPID controller	Controller equation	Order range
$(PI^{\lambda} D^{\mu})$	$C(s)_{FOPID} = \left(K_P + \frac{K_I}{s^{\lambda}} + K_D s^{\mu} \right) \cdot E(s)$	$\lambda, \mu \in [0, 2]$
$(PID)^N$	$C(s)_{(PID)^N} = \frac{1}{s^N} \left(K_P + \frac{K_I}{s} + K_D s \right) \cdot E(s)$	$N \in [0, 1]$
IMCFOPID	$C(s)_{IMCFOPID} = \left(K_P + \frac{K_I}{s^{\alpha}} + \frac{K_D s^{\alpha}}{\gamma s^{\alpha} + 1} \right) \cdot E(s)$	$\alpha \in [0,1]$
$(PI)^{\lambda}$-$(PD)^{\mu}$	$C(s)_{(PI)^{\lambda}} = \left(K_P + \frac{K_I}{s} \right)^{\lambda} \cdot E(s)$ $C(s)_{(PD)^{\mu}} = (K_P + K_D s)^{\mu} \cdot E(s)$	$\lambda \in [0, 2]$ $\mu \in [0, 2]$
Modified FOPI	$C(s)_{MFOPI} = \left(K_P \cdot \frac{1 + T_D s}{s^{\alpha}} \cdot \frac{s + \eta}{s} \right) \cdot E(s)$	$\alpha \in [0, 1]$
Modified FOPID	$C(s)_{MFOPID} = \left(K_C \cdot \frac{(1 + T_D s^{\lambda})^2}{s^{\lambda}} \right) \cdot E(s)$	$\lambda \in [0, 1]$

Different types of FOPID controllers along with the above three are listed in Table 10.1.

10.3.2.3 *Effects of Fractional-Orders in Controller Performance*

In FOPID controller, the integral and derivative terms are fractionally ordered. In case of FO integration, the order of the integral defines the particular area under the error curve that should be taken into account during controlling. So, the fractional integral term works with the selective memory of the past error values. Integral action improves steady-state behaviour of the controller. Now, the physical significance of derivative action is to predict the error of future output. The fractional-order of derivative selects the prediction horizon. This is achieved by linear or nonlinear extrapolation of error curve [27]. Derivative action also improves the overall system stability in a closed loop. If the error signal is constant, the

Table 10.2 Effects of fractional-order PID actions.

Action	Domain	Effect on	Effect of fractional order
Integral	Time	Controller output	Nonlinearly increases or decreases according to the sign of error with time
		Steady-state error	Eliminates it perfectly
		Proportional Bias	Rate of change and the shape of resetting can be controlled
	Frequency	Magnitude plot	Increments magnitude curve with the slope of -20λ dB/dec in a constant manner
		Phase plot	Constantly lags the phase curve by $(\pi\lambda)/2$ rad
Derivative	Time	Constant error signal	Responds to it and can be used in isolated way
		Prediction horizon	Prediction horizon is set by using linear and nonlinear extrapolation over error curve through current error point; The extrapolating curve's or straight line's slope is μ
	Frequency	Magnitude plot	Increments magnitude curve with the slope of 20λ dB/dec in a constant manner
		Phase plot	Constantly leads the phase curve by $(\pi\lambda)/2$ rad
		Filter design	Low-pass filter of order μ is possible
			Depending on the value of μ frequency responses between consecutive integer order filters can be achieved

classical derivative action fails to respond, but fractional-order derivative action can respond to it as a fractional-order derivative of any constant that is not 0. Effects of the fractional order PID controller's integral and derivative actions are listed in Table 10.2.

10.4 Modelling of Robotic Manipulator

This chapter focuses on controlling of robotic manipulator system by means of Fuzzy Fractional-order PID (FFOPID) control. In this section, modelling of a single-link robotic manipulator is elaborated.

10.4.1 Modelling of Single-Link Manipulator

Single-link manipulators are a very basic form of robotic manipulators. It contains one arm, one revolute joint, one servo motor at the joint, and a fixed base, as shown in Figure 10.2. Thus, a single-link manipulators possess one-degree of freedom. The single-link manipulator model will be derived by modelling the dynamics of the servo motor at the joint and the rigid arm separately. The overall transfer function is obtained for the overall system. The Newton-Lagrange approach is used here.

10.4.2 Modelling of Dynamics of Servo Motor

The servo motor provides the mechanical energy required for manipulator rotation. From Figure 10.1, it can be noticed that the armature circuit of a servo motor consists of resistance R_a and inductance L_a. The mechanical load (here the manipulator arm) is connected across M. The current in the circuit is denoted as i_a. Voltage V is applied across the circuit.

According to KVL, the following equation can be written for the armature circuit in Figure 10.2:

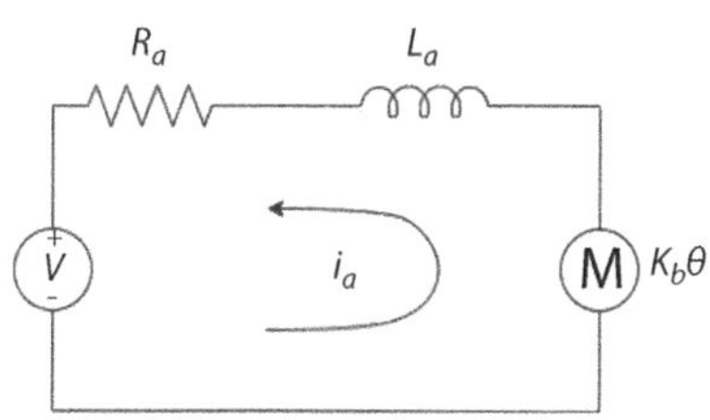

Figure 10.1 Servo motor armature circuit.

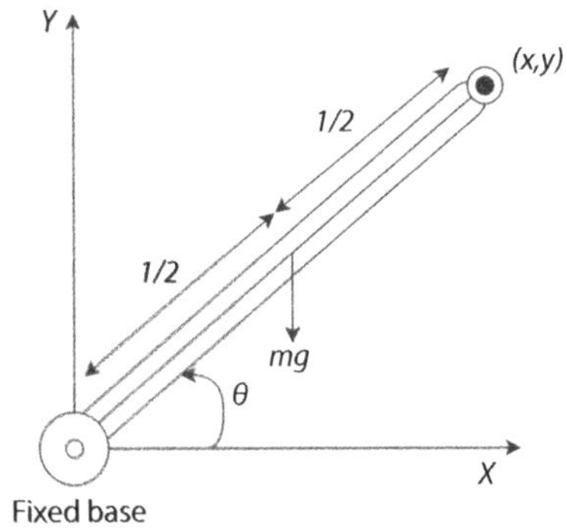

Figure 10.2 Single-link manipulator mechanism.

$$V = L_a \frac{di_a}{dt} + K_b\dot{\theta} + R_a i_a \tag{10.19}$$

Here, K_b is the back-emf constant of the servo motor and θ is the angular displacement of the motor shaft.

Applying the Laplace transform on Equation 10.19:

$$\begin{aligned} V(s) &= (L_a s + R_a) I_a(s) + K_b s \cdot \theta(s) \\ \Rightarrow I_a(s) &= \frac{V(s) - K_b s \cdot \theta(s)}{(L_a s + R_a)} \end{aligned} \tag{10.20}$$

Now, the Laplace transforming torque equation of the servo motor:

$$T(s) = K_T \cdot I_a(s) \tag{10.21}$$

Here, *T(s)* is the Laplace transform of torque produced by the motor and K_T is the torque constant of the motor.

10.4.3 Modelling of Manipulator Dynamics

From Figure 10.2, the coordinate of the manipulator tooltip *(x,y)* can be written as:

$$\begin{aligned} x &= \frac{l}{2}\cos\theta \\ y &= \frac{l}{2}\sin\theta \end{aligned} \tag{10.22}$$

Here, l is manipulator arm length.

Overall velocity of the arm across its tooltip is:

$$\upsilon = \sqrt{(\dot{x})^2 + (\dot{y})^2} = \sqrt{\left(-\frac{l}{2}\dot{\theta}\sin\theta\right)^2 + \left(\frac{l}{2}\dot{\theta}\cos\theta\right)^2} = \frac{l}{2}\dot{\theta} \quad (10.23)$$

The kinetic energy is given by:

$$E_K = \frac{1}{2}(m\upsilon^2 + I\dot{\theta}^2) = \frac{1}{6}ml^2\dot{\theta}^2 \quad (10.24)$$

Here, m is the mass of manipulator arm and I is the inertia of the arm.

The potential energy is given by:

$$E_P = \frac{1}{2}mgl\sin\theta \quad (10.25)$$

From Lagrange's Equation:

$$\frac{d}{dt}\left(\frac{\partial \mathcal{L}}{\partial \dot{\theta}}\right) - \frac{\partial \mathcal{L}}{\partial \theta} = T$$

$$\Rightarrow \frac{1}{3}ml^2\ddot{\theta} + \frac{1}{2}mlg\cos\theta = T \quad (10.26)$$

Here, T is the torque of the manipulator arm, $\mathcal{L} = E_K - E_P$.

Now, considering θ is very small, it can be assumed that $\cos\theta \approx 1$. Also, considering the damping effect of the servo motor, Equation 10.26 becomes

$$\frac{1}{3}ml^2\ddot{\theta} + \frac{1}{2}mgl\theta + b\dot{\theta} = T$$

$$\Rightarrow 2ml^2\ddot{\theta} + 6b\dot{\theta} + 3mgl\theta = 6T$$

$$\Rightarrow J\ddot{\theta} + 6b\dot{\theta} + h\theta = 6T, \qquad J = 2ml^2, h = 3mgl \quad (10.27)$$

Here, $b\dot{\theta}$ represents the damping of the motor.

Applying the Laplace transform on Equation 10.27 and from Equation 10.21:

$$\theta(s)(Js^2+6bs+h)=6K_T I_a(s)$$
$$\Rightarrow I_a(s)=\frac{\theta(s)(Js^2+6bs+h)}{6K_T} \tag{10.28}$$

Comparing Equations 10.28 and 10.20, the transfer function of the overall system is obtained:

$$G(s)=\frac{\theta(s)}{V(s)}=\frac{6K_T}{(Js^2+6bs+h)(L_a s+R_a)+s\cdot K_T K_b} \tag{10.29}$$

The transfer function described in Equation 10.29 will be used for controlling in the later part of this chapter.

10.5 Proposed Design of Fuzzy Fractional-Order PID Controller

The current section elaborates on the design of the proposed fuzzy logic based FOPID controller for angular position control of robotic manipulators.

10.5.1 Implementation of Fuzzy Logic for Gain Scheduling of FOPID

The gains of conventional PID controllers are arbitrary positive constants which should be tuned for obtaining the desired response. There are many procedures available for tuning these gains like conventional Zeigler-Nichols, Cohen-Coon Method, optimal control techniques, neural network based tuning, and many more. But, in case of industrial application, most of the tuning techniques do not give desirable performance, as for using those tuning techniques complete information of the process model should be known. For solving this problem different model identification techniques have been developed. These techniques are quite expensive in terms of computational means. As a solution to that problem, a Fuzzy Logic Controller (FLC) can be introduced. FLC is based on set of rules. Depending on that rule base a parameter can be tuned. So, for designing an FLC for tuning a parameter only the characteristics of that parameter are required to be known. Also, FLCs are computationally cheap. Implementing

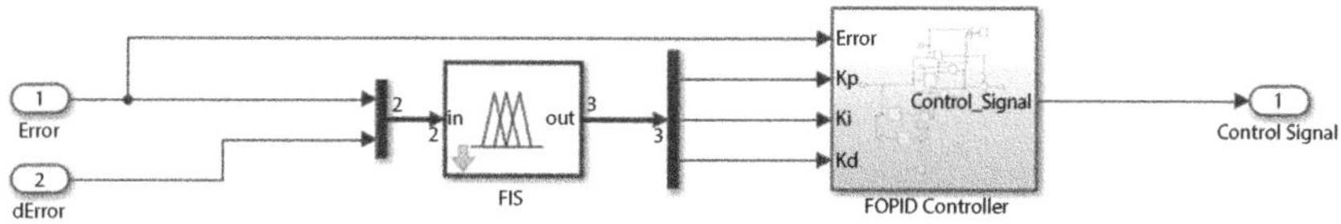

Figure 10.3 Structure of FIS and FOPID controller.

an FLC in industry is quite feasible. In this sub-section designing of an FIS in Figure 10.3 for self-tuning K_p, K_i, and K_d gains of fractional-order PID controllers is discussed.

10.5.2 Structure of the Proposed Fuzzy Inference System

For achieving self-tuning goals, a Mamdani type FIS is designed. Refer to the figure above for the rest of the sub-section.

Linguistic Variables: For all inputs and outputs, five linguistic variables are assigned which are summarized in Table 10.3.

10.5.2.1 Inputs

There are two inputs of the FIS. The first one is the error *(e)*, which is measured by subtracting controlled process output from set-point. The second one is the rate of change in error *(de)*, which is measured by differentiating the instantaneous error value at that point of time. Ranges of these inputs are taken as $e \in [-20, 20]$ and $de \in [-20, 20]$. For fuzzification of both inputs, trapezoidal membership functions are used for NB & PB and Triangular membership functions are implemented for rest of

Table 10.3 List of linguistic variables for designing FIS.

Linguistic variable	Interpretation
NB	Negative Big
NS	Negative Small
ZO	Zero
PS	Positive Small
PB	Positive Big

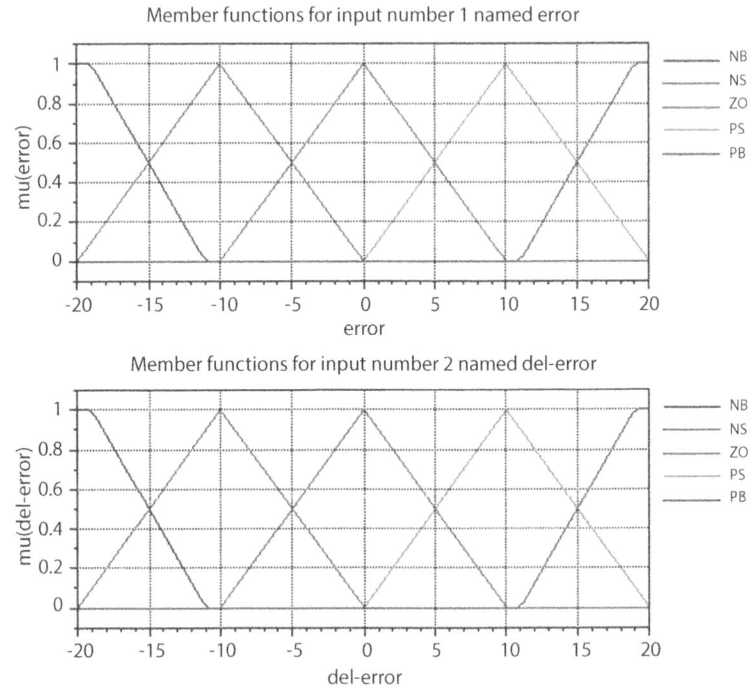

Figure 10.4 Membership functions plot for inputs of designed FIS.

the linguistic variables. Levels for linguistic variables are uniformly distributed. Membership functions for *e* (error) and *de* (del-error) are shown in Figure 10.4.

10.5.2.2 Outputs

The FIS has three outputs. Each for giving K_p, K_i, and K_d values respectively. Ranges of these outputs are taken as $K_p \in [0, 100]$, $K_i \in [0, 200]$ and $K_d \in [0, 05]$. These ranges can be modified as per application. Throughout the chapter these ranges have been used for simulation purposes. Gaussian Membership functions are considered for all the linguistic variables of both outputs. Levels for linguistic variables are uniformly distributed too. Membership functions for K_p, K_i, and K_d are shown in Figure 10.5.

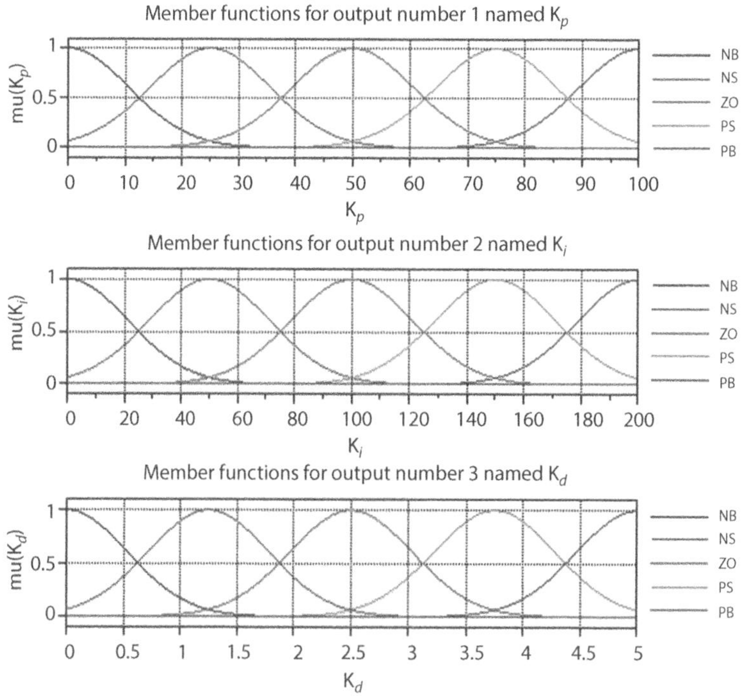

Figure 10.5 Membership functions plot for outputs of designed FIS.

10.5.2.3 Rule-Base

Three different rule-bases are designed [28]. Individual PID gain characteristics are listed in Table 10.4 and rule bases for determining PID controller gains (Kp, Ki & Kd) are listed in Table 10.5–Table 10.7 respectively.

Each rule in the rule-base is designed in the following form with two antecedents and three consequents:

Table 10.4 Characteristics of PID gains.

Gain	Variation	Rise time	Overshoot	Settling time	Steady-state error	Stability
K_p	Increase	Decreases	Increases	Slightly increases	Decreases	Degrades
K_i	Increase	Decreases	Increases	Increases	Eliminates	Degrades
K_d	Increase	Slightly decreases	Decreases	Decreases	No effect	Improves

Table 10.5 Rule-base for determining K_p.

e de	**NB**	**NS**	**ZO**	**PS**	**PB**
NB	NB	NB	NB	NS	ZO
NS	NB	NS	NS	NS	ZO
ZO	NB	NS	ZO	PS	PB
PS	ZO	PS	PS	PS	PB
PB	ZO	PS	PB	PB	PB

Table 10.6 Rule-base for determining K_i.

e de	**NB**	**NS**	**ZO**	**PS**	**PB**
NB	PB	PB	PB	NS	NB
NS	PB	PS	PS	ZO	NB
ZO	PS	PS	ZO	NS	NB
PS	ZO	PS	NS	NS	NB
PB	ZO	NS	NB	NB	NB

Table 10.7 Rule-base for determining K_d.

e de	**NB**	**NS**	**ZO**	**PS**	**PB**
NB	NB	NB	NB	PS	PB
NS	NB	NS	NS	ZO	PB
ZO	NS	NS	ZO	PS	PB
PS	ZO	NS	PS	PS	PB
PB	ZO	PS	PB	PB	PB

$$R^i\text{: IF } e \text{ is } A_1^j \text{ and } de \text{ is } A_2^k\text{, THEN } K_p \text{ is } C_i^l \text{ and } K_i \text{ is } C_i^m \text{ and } K_d \text{ is } C_i^n.$$

$$1 \le j,k,l,m,n \le 5$$

10.5.2.4 Inference and Defuzzification Technique

The decision-making for obtaining output is done using the max-min inference technique and defuzzification of the output fuzzy variable is done using the Center-of-Gravity (COG) approach. Input-Output surface of the designed FIS is shown in Figure 10.6.

Rule-viewer example given in Figure 10.7.

$$ITAE = \int_0^\infty t \cdot |e(t)| \cdot dt \tag{10.30}$$

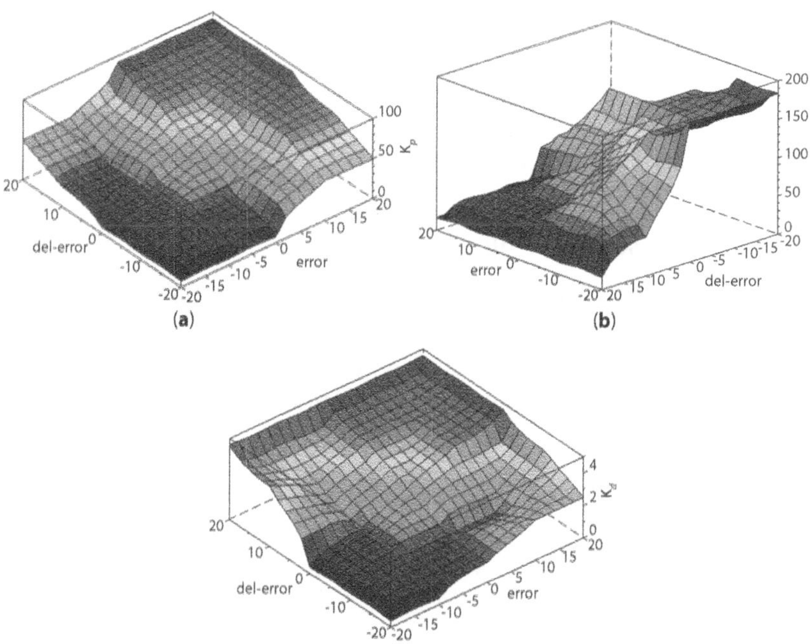

Figure 10.6 (a) Fuzzy IO surface for K_p, (b) Fuzzy IO surface for K_i, (c) Fuzzy IO surface for K_d.

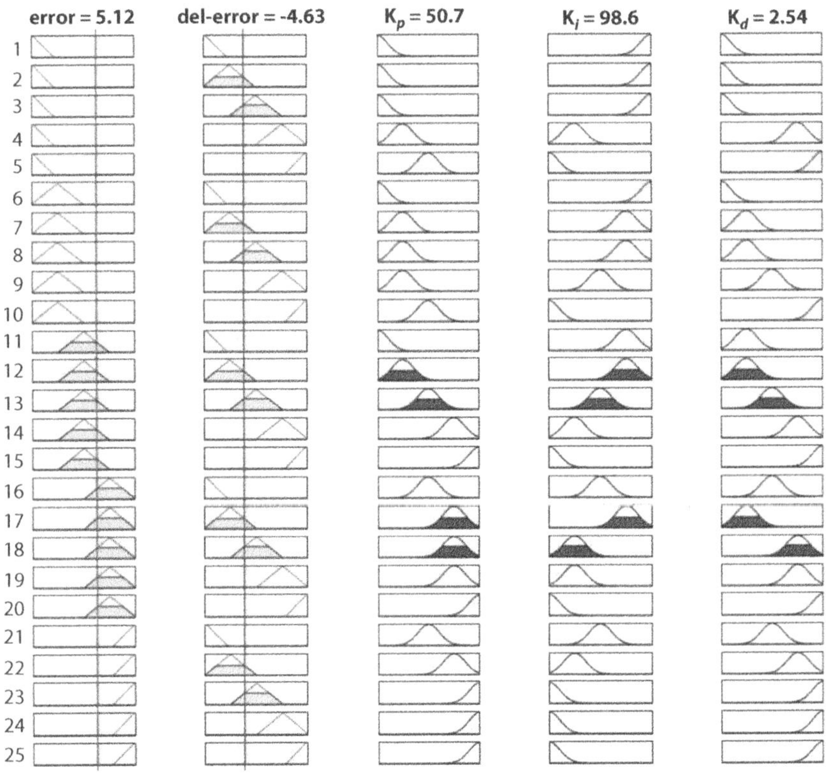

Figure 10.7 Fuzzy rule-viewer example.

10.5.3 Proposed Controller Structure

Proposed controller structure connecting the FIS, FOPID controller and the plant model is shown in Figure 10.8.

Comparative Performance Evaluation of three models of FOPID controllers, e.g., Podlubny's model, M. Tenoutit's model, and IMC based

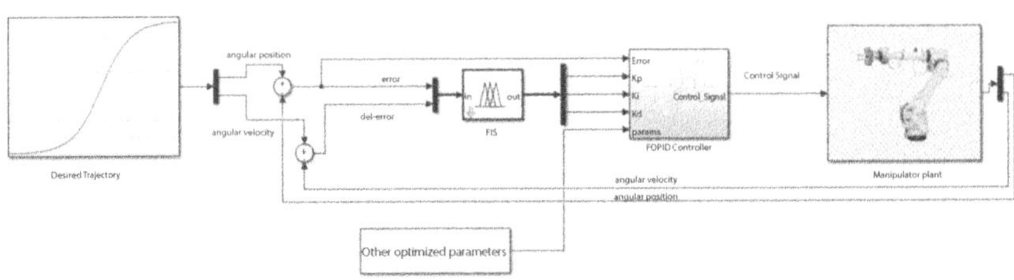

Figure 10.8 Overall Fuzzy FOPID controller-in-loop structure.

FOPID model, along with a conventional parallel PID model with fuzzy gain scheduling is studied. As a plant model, a manipulator with a single link is chosen. Based on the results, the best controller model is selected based upon the simulation based results.

10.6 Simulation Study of Proposed FFOPID Controller

The simulation study is performed using MATLAB-SIMULINK in this chapter. A PC with Intel i3 processor and 8GB of RAM is used. FOMCON toolbox and Optimization toolboxes are used respectively for fractional-order controller synthesis and an optimization problem solution for the same [29]. A unit step signal is used as a reference signal. A comparative study of controller performance is carried out for the three different FFOPID controller models proposed here with a Fuzzy PID controller. Based upon different performance metric scores, the best controller model is chosen for a single link manipulator model. The simulink model block diagram is given in Figure 10.9. Here, only the FOPID controller type is updated for recording its performance.

Single-link manipulator model parameters are given in Table 10.8.

Table 10.9 summarizes the optimum FOPID parameters used for this simulation study.

From Figure 10.10, it is evident that the Fuzzy logic based Podlubny's FOPID model and IMC based FOPID model outperforms the Fuzzy logic based PID and M. Tenoutit's FOPID models. Figure 10.11 shows that the error converges towards null value faster in the case of a Fuzzy logic based Podlubny's and IMC-FOPID model. Control actions are shown in

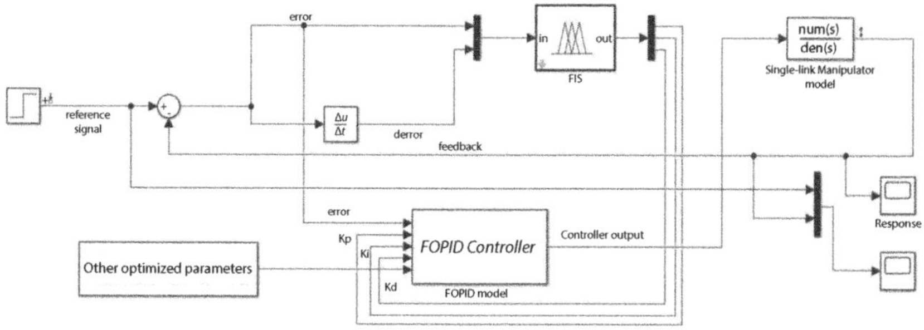

Figure 10.9 Simulink block diagram for comparative performance evaluation study of FFOPID controllers.

Table 10.8 Specification of single-link manipulator system.

Parameter	Value
R_a	10 Ohms
L_a	0.50 H
m	1 kg
b	5 N.m.s/rad
K_T	140 N.m/A
K_b	140 V.s/rad
l	2.07 m
J	8.60
p	126.42

Table 10.9 Optimized values of FOPID model parameters.

FOPID model type	Variables for optimization	Bounds		Fitness function	Optimum value	Best fitness score
		Lower	Upper			
Podlubny's Model	λ, μ	[0.1 0.1]	[1 1]	ITAE	$\lambda = 0.1016$ $\mu = 0.427$	0.00810091
M. Tenoutit's Model	N	[0.001]	[1]		$N = 0.0011$	0.0437799
IMC-FOPID	α, γ	[0.01 0.001]	[1 1]		$\alpha = 0.053$ $\gamma = 0.007$	0.00673547

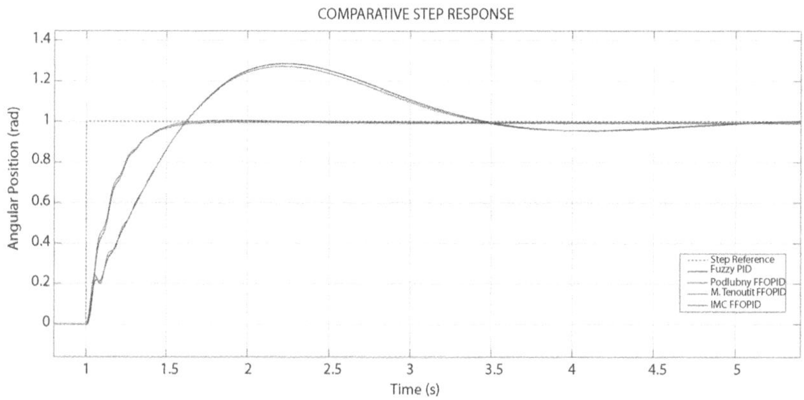

Figure 10.10 Step response of different FFOPID controllers.

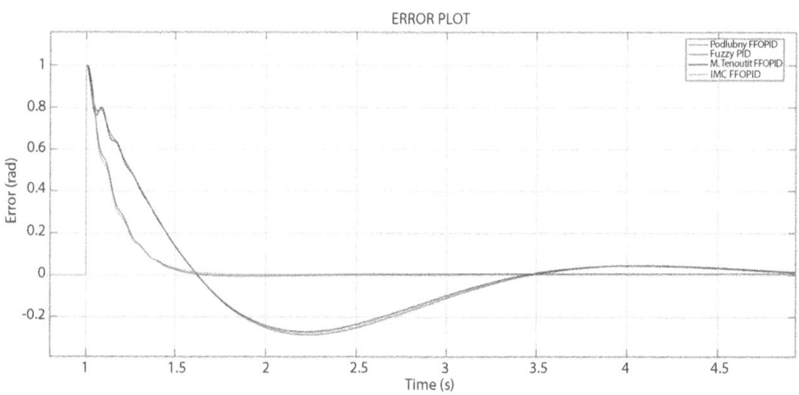

Figure 10.11 Evolution of error with time for different FFOPID controllers.

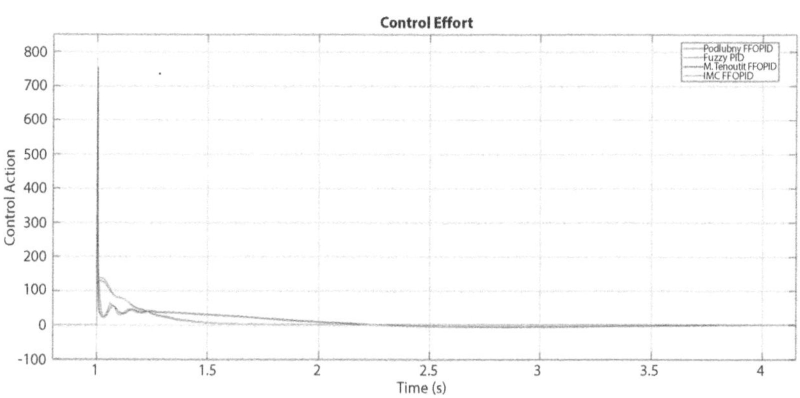

Figure 10.12 Comparative plot of control action.

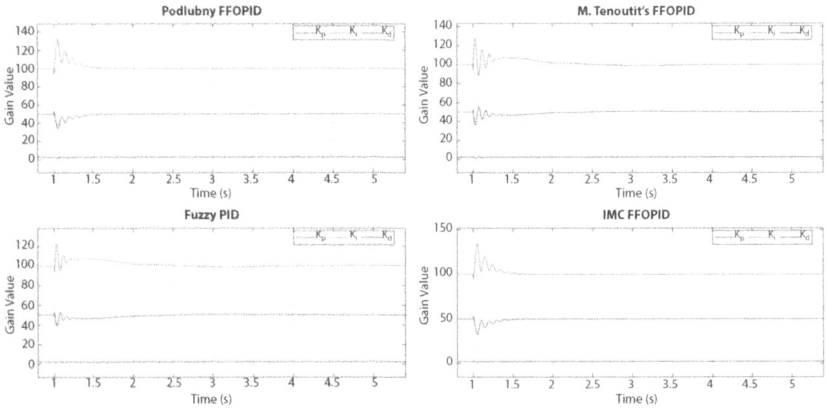

Figure 10.13 PID gain variation plot for different FFOPID controllers.

Figure 10.12. It can be noticed from Figure 10.13 that q variation in PID gain values is less in Fuzzy PID models followed by Podlubny's model, but in the case of Podlubny's model and IMC-FOPID model PID gain variation settles very quickly.

10.6.1 Analysis of Step Response

In Table 10.10, the obtained time domain parameters and stability margins are summarized. From the step response analysis it can be comprehended that a Fuzzy logic based Podlubny's FOPID model and IMC-FOPID model surpass the other two controllers with a very significant performance margin. The fuzzy logic based IMC-FOPID model outperforms the conventional one except the settling time value. The obtained IAE and ITAE score for the Fuzzy IMC-FOPID controller is a little bit higher than Podlubny's model and the ISE score is lesser for this model.

Furthermore, this Fuzzy IMC-FOPID controller eliminates overshoot satisfactorily with a very smooth and minimal control action which is very necessary in case of manipulator robotics application as overshoots may cause various problems in manipulator control like excess torque production at joints, setting joint angles outside of joint-space, etc. Also, settling time for this controller is much less, almost half of a second which is very necessary for high precision fast industrial manipulators. The stability margin for Fuzzy IMC-FOPID is better than other controllers in the above table. Hence, the performance of Fuzzy IMC-FOPID has shown better performance than other controllers mentioned in the table above.

Table 10.10 Performance comparison of proposed controller.

FFOPID model	Time domain				Frequency domain		Error metric			Closed-loop stability
	Rise time (s)	Overshoot (%)	Settling time (s)	Steady-state value (rad)	Gain margin (dB)	Phase margin (°)	ISE	IAE	ITAE	
Podlubny's Model	0.321	0.457	0.538	1.0	10.5	-180	0.08766	0.204	0.4826	Stable
M. Tenoutit's Model	0.516	26.5	3.79		18.4	-64	0.2177	0.6084	1.262	Stable
IMC-FOPID	0.319	0.0	0.579		9.75	-180	0.08443	0.2161	0.5565	Stable
Fuzzy PID	0.514	26.2	3.79		12.2	-63	0.2285	0.6304	1.317	Stable

10.7 Conclusion

In this chapter, a Fuzzy logic based partially-optimized FOPID controller is developed mainly focusing on controlling robot manipulators. Before going into developing the proposed controller. brief introductions to Fuzzy Logic Systems, Fractional-order calculus, and different types of FOPID controllers are given. Also, explicit mathematical modelling of a single-link system is also done. In the later part of the chapter, the proposed controller model is evaluated by performing angular position control on a single-link manipulator model and the best case model is identified. The proposed controller model gave the desired response accurately and proves it to be usable in controlling industry-grade manipulator models. Future work for developing this controller by using a Takagi-Sugeno type FIS and testing its efficacy in controlling multilink manipulators may be considered.

References

1. Kocian, J., Koziorek, J., & Pokorny, M. (2011). An Approach to PLC-Based Fuzzy Expert System in PID Control. *Informatics Engineering and Information Science*, 195–205.
2. Abdollah Nezhad, Q., Palizvan Zand, J., & Shah Hoseini, S. (2013). An Investigation on Fuzzy Logic Controllers (Takagi-Sugeno & Mamdani) in Inverse Pendulum System. *International Journal of Fuzzy Logic Systems*, *3*(3), 1–14.
3. Precup, R. E., Tomescu, M. L., Preitl, S., Petriu, E. M., & Dragoş, C. A. (2011). Stability Analysis of Fuzzy Logic Control Systems for a Class of Nonlinear SISO Discrete-Time Systems. *IFAC Proceedings Volumes*, *44*(1), 13612–13617.
4. Karnik, N., Mendel, J., & Qilian Liang. (1999). Type-2 fuzzy logic systems. *IEEE Transactions on Fuzzy Systems*, *7*(6), 643–658. https://doi.org/10.1109/91.811231
5. O'Brien, R. T., & Howe, J. M. (2008). Optimal PID controller design using standard optimal control techniques. *2008 American Control Conference*.
6. Zhang, Y., Yu, X., Bi, M., & Xu, S. (2011). An adaptive neural PID controller for torque control of hybrid electric vehicle. *6th International Conference on Computer Science & Education*.
7. Caponetto, R., Dongola, G., Fortuna, L., & Petras, I. (2010). *Fractional order systems: modeling and control applications. World Scientific Nonlinear Science Series a.*
8. Bingi, K., Ibrahim, R., Karsiti, M. N., Hassan, S. M., & Harindran, V. R. (2019). Scilab Based Toolbox for Fractional-order Chaotic Systems. *Fractional-Order Systems and PID Controllers*, 213–245.

9. Shah, P., & Agashe, S. (2016). Review of fractional PID controller. *Mechatronics, 38*, 29–41.
10. Tavazoei, M. (2014). Time response analysis of fractional-order control systems: A survey on recent results. *Fractional Calculus and Applied Analysis, 17*(2), 440–461.
11. Petras, I., & Kover-Dorco, M. (2016). An effective algorithm for implementation of non-linear fractional-order controller on PLC. *17th International Carpathian Control Conference.*
12. Merrikh, B. F., & Mirebrahimi, N. (2011). Introduction to the nonlinear $PI^{\lambda}D^{\mu}$ control. *IEEE International Conference on Control System, Computing and Engineering.*
13. Kumar, A., Gaidhane, P. J., & Kumar, V. (2017). A nonlinear fractional order PID controller applied to redundant robot manipulator. *6th International Conference on Computer Applications in Electrical Engineering-Recent Advances.*
14. Zadeh, L. (1965). Fuzzy sets. *Information and Control, 8*(3), 338–353.
15. Shaw, I. S. (1998). Generic Structure of Fuzzy Controllers. *Fuzzy Control of Industrial Systems*, 55–66.
16. Tepljakov, A. (2018). *Fractional-order Modeling and Control of Dynamic Systems.* Springer.
17. Chen, Y., Petras, I., & Xue, D. (2009). Fractional order control - A tutorial. *American Control Conference.*
18. Machado, J. T., Kiryakova, V., & Mainardi, F. (2011). Recent history of fractional calculus. *Communications in Nonlinear Science and Numerical Simulation, 16*(3), 1140–1153.
19. Monje, C. A., Chen, Y., Vinagre, B. M., Xue, D., & Feliu, B. V. (2014). *Fractional-order Systems and Controls: Fundamentals and Applications.* Springer.
20. Pan, I., & Das, S. (2013). Applied Fractional Calculus for Computational Intelligence Researchers. *Intelligent Fractional Order Systems and Control*, 9–61.
21. Senol, B., & Yeroglu, C. (2014). Filter approximation and model reduction comparison for fractional order systems. *ICFDA'14 International Conference on Fractional Differentiation and Its Applications 2014.*
22. El-Khazali, R., Batiha, I. M., & Momani, S. (2019). Approximation of Fractional-Order Operators. *Springer Proceedings in Mathematics & Statistics*, 121–151.
23. Podlubny, I. (1999). Fractional-order systems and $PI^{\lambda}D^{\mu}$-controllers. *IEEE Transactions on Automatic Control, 44*(1), 208–214.
24. Luo, Y., & Chen, Y. (2009). Fractional order [proportional derivative] controller for a class of fractional order systems. *Automatica, 45*(10), 2446–2450.
25. Tavakoli-Kakhki, M., & Haeri, M. (2011). Fractional order model reduction approach based on retention of the dominant dynamics: Application in IMC based tuning of FOPI and FOPID controllers. *ISA Transactions, 50*(3), 432–442.

26. Feliu, B. V., Pérez, R. R., & Rodríguez, L. S. (2007). Fractional robust control of main irrigation canals with variable dynamic parameters. *Control Engineering Practice, 15*(6), 673–686.
27. Tejado, I., Vinagre, B., Traver, J., Prieto-Arranz, J., & Nuevo-Gallardo, C. (2019). Back to Basics: Meaning of the Parameters of Fractional Order PID Controllers. *Mathematics, 7*(6), 530.
28. Sinthipsomboon, K., Hunsacharoonroj, I., Khedari, J., Po-ngaen, W., & Pratumsuw, P. (2012). A Hybrid of Fuzzy and Fuzzy Self-Tuning PID Controller for Servo Electro-Hydraulic System. *Fuzzy Controllers-Recent Advances in Theory and Applications*. Published. https://doi.org/10.5772/48614
29. Tepljakov, A. (2017). FOMCON: Fractional-Order Modeling and Control Toolbox. *Fractional-Order Modeling and Control of Dynamic Systems*, 107–129.

11

Prototype Development of an Electromagnetic Levitation System for Maglev Vehicle

J. Kundu* **and A. Choumal**

Department of Electrical Engineering, UCET, Bikaner, Rajasthan, India

Abstract

This chapter presents a brief study, analysis, validation and control of an electromagnetic attraction type levitation set-up where a flat plate has been successfully levitated at a desired operating point. Electromagnetic finite analysis has been done for this asymmetric shaped object. A good agreement between simulated and practical parameter evaluation leads to a precise and robust set-up. Modeling and stability analysis has been done accurately. Finally, a controller has been successfully implemented on the lab developed prototype. As the plate has an asymmetric shape, the complexities arising in coil design analysis and implementation of controllers have been focused on this work broadly.

Keywords: Magnetic levitation, lead compensator, FEM analysis, electromagnetic coil

11.1 Introduction

Maglev (*magnetic levitation*) popularly refers to a system of transport that uses magnetic levitation to suspend, guide, and drive vehicles with the help of magnets without the use of mechanical elements such as wheels, axels, and bearings [1–3]. The many other applications of attraction type electromagnetic levitation are in wind tunnels, magnetic bearings, wafer transportation, vibration isolation in sensitive equipment, suspension of molten metal in induction furnaces, etc. [4–8]. There are also applications in the semiconductor and biomedical industries. With the help of maglev,

**Corresponding author*: janardan.netaji@gmail.com

Vipin Chandra Pal, Suman Lata Tripathi, and Souvik Ganguli (eds.) Industrial Control Systems, (231–254) © 2024 Scrivener Publishing LLC

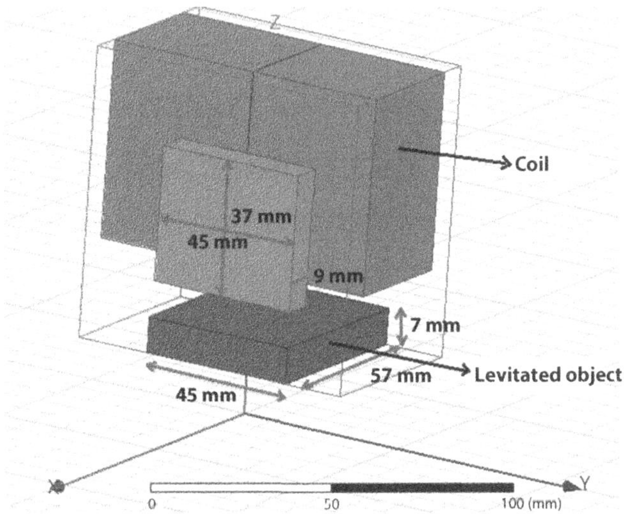

Figure 11.1 Mechanical dimensions of levitation system.

a vehicle is levitated just above the guide way to create both lift and thrust using magnets or magnetic fields. In the upcoming generation, High-speed maglev vehicles can bring striking improvements in human civilization in terms of its comfortableness and saving of time in a surprising manner [19, 20]. In propulsion applications, the energy required for suspension is usually a much smaller percentage of the total energy consumption [16].

Use of levitation in applications, e.g., magnetic bearings, magnetically lifted trains, etc. has drawn huge attention throughout worldwide research [2, 25]. The advantage is frictionless, which further leads to lower heating of the moving parts which results in a higher running efficiency. It increases energy efficiency, running efficiency, reduces long term maintenance costs, and prolongs the life of the overall systems. The literature which is available in the field of magnetic levitation is mainly simple single magnet-based suspension systems [26, 27]. The conventional design approach uses different methods to deal with non-linearity. In [25], Banerjee *et al.* have discussed controller design for gap control of electronic magnetic levitation systems using an optimization technique. However, a mechanical block has been used to resist the movement in the transverse direction. Wong *et al.* [9] has presented the design and control strategies of the attraction type electromagnetic levitation system. In [4], Moon *et al.* have discussed superconducting magnet based levitation, application to magnetic bearings, and magnetic vehicles, but does not address analytical modeling in any of the above literature on electromagnetic levitation. In [5], Banerjee *et al.* have highlighted a method for the development of electromagnetically suspended vehicles using an unifying approach. Venkataratnam *et al.* has presented a theory and performance predictions of the rotor position in a switched reluctance using analytical modeling [28]. Jinhui *et al.* have discussed the control of maglev with a virtual energy harvester [12]. A new maglev system for magnetically lifted carrier systems has been presented by Mofushita *et al.* in [13]. Ohatsuka *et al.* has discussed the high speed ground transport project by superconducting magnets in Japan [10]. Ma *et al.* have presented an integrated levitation system [11].

Rogg *et al.* [6] have discussed a general study on possible levitation applications and development trends in magnetic levitation. However, no study on disturbances along the x-direction is highlighted. The method of zero power control in electromagnetic levitation systems has been discussed in [22–24]. The controller design and implementation for a magnetic suspension system has been presented by Shiao *et al.* [21], but no comparative study has been performed for the various levitation prototypes. Again, in [17], Moriyama *et al.* have discussed AC magnetic levitation using a differential feedback linear amplifier. However, no research has been presented on the validity of analytical, practical, and simulation results. Tsukamoto *et al.* [18] have discussed the new magnetic levitation systems with AC magnets. However, no effect of change of dimensions and asymmetric shape of the levitated object has been studied. In [19], Yamaguchi *et al.* discussed electromagnetic levitation systems with laminated slot less rails, but no research has been conducted on coil design for objects of different shapes. A unique method of controller design in levitation systems has been highlighted in [15, 16]. However, no set-up upgradation has been discussed. Though in [14] digital controller implementation for magnetic suspension system has been talked about, no FEM modeling has been addressed in the above literature. Jayawant has presented a brief review of the literature on electromagnetic levitation techniques in [2].

11.1.1 Maglev Transportation

Engineering aspects of some of the components of the system are presented for evaluation of suitability for low speed systems. Transrapid transport systems can have speeds up to 500 *Km/h*. It is characterized by non-contacting. Oliveira *et al.* have discussed the digital implementation of

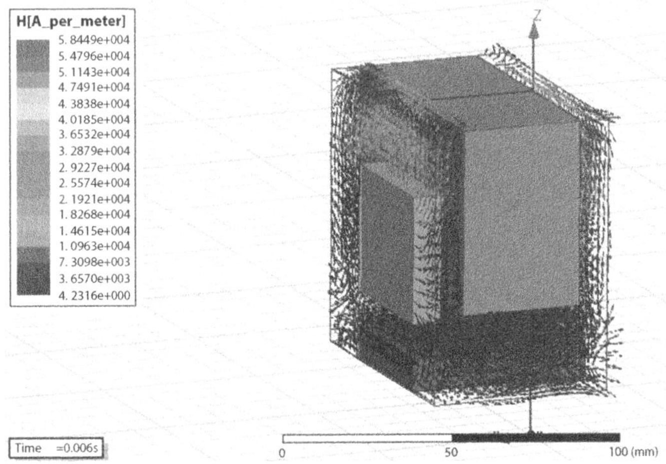

Figure 11.2 3D view of model.

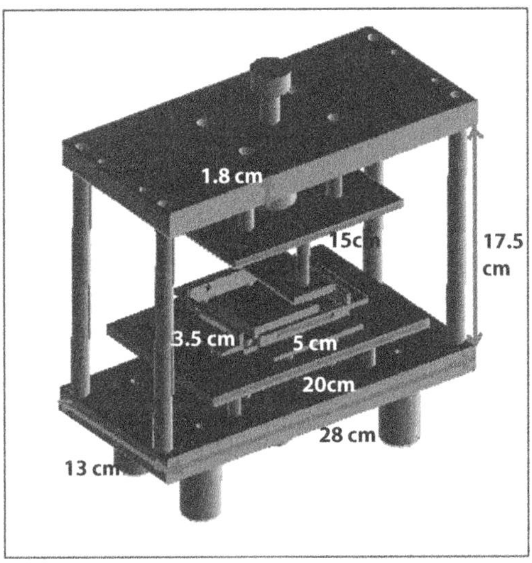

Figure 11.3 AutoCAD drawing of fabricated aluminum made mechanical structure.

magnetic suspension systems for laboratory tests in [14]. Electromagnetic suspension and levitation technology has been addressed in [2]. Superconducting in axial bearings for active induction devices are further discussed by Nikolsky *et al.* in [29]. Superconducting magnetic bearings using YBCO bulk block were also introduced in [30]. Development and optimization of TRANSRAPID vehicles has been carried out successfully

Figure 11.4 Photograph of proposed levitation system.

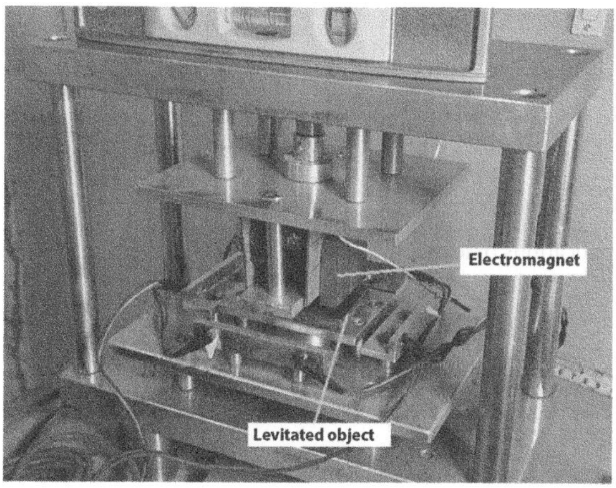

Figure 11.5 Full prototype of levitation system.

in Emsland. Results of vehicle performance analysis, linear motors, and guideways has been analysed. These are verified by partial data and real test results. For further technological developments, the Federal Republic of Germany is preparing to implement a high speed transport system. Proof of operational readiness is a priority as a prerequisite for concrete track planning. This transport system will initially operate on the FRG's main traffic route and as a feeder service from the airport to downtown. Maglev technology (shown in Figures 11.1-11.6) holds great promise for the future. It is cheaper, faster, and safer and has the potential for greener transportation than we have today (Figure 11.6, Figure 11.7, Figure 11.8, Figure 11.9 and Figure 11.10). The benefits are very hard to ignore. So far, only one commercial Maglev train (Figures 11.11 and 11.12) is being used and it has already exceeded that which preceded it.

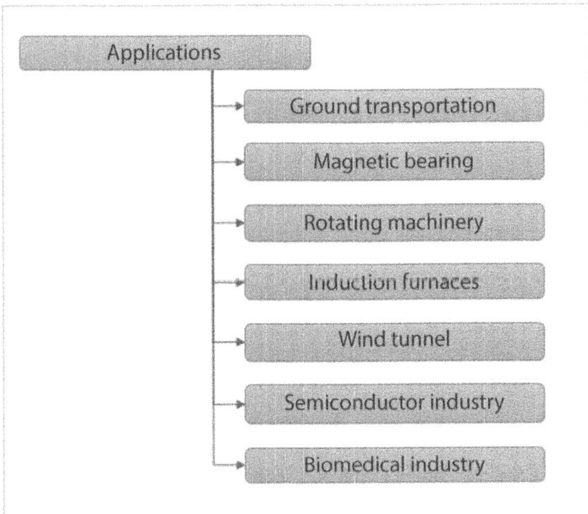

Figure 11.6 Applications of levitation.

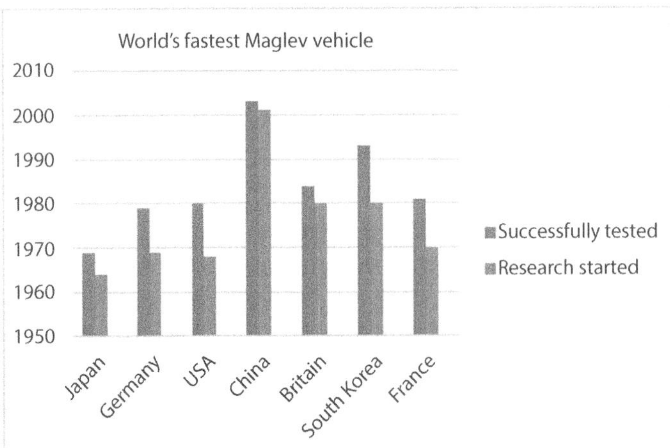

Figure 11.7 Global scenario of maglev transport.

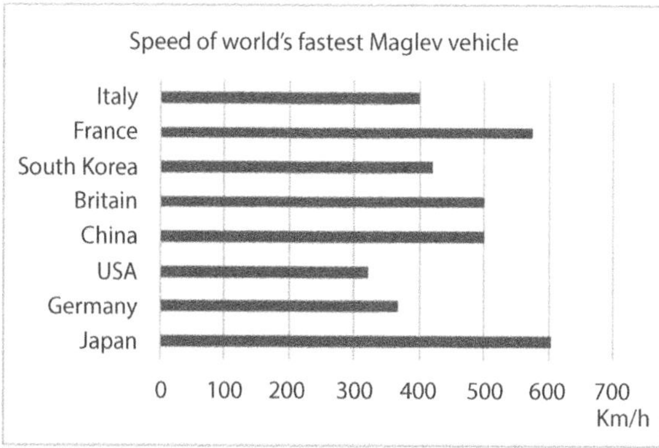

Figure 11.8 Speed of world's fastest maglev vehicle.

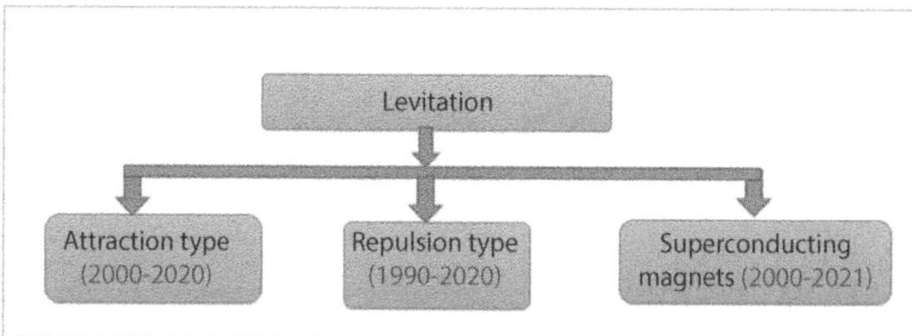

Figure 11.9 Classifications of levitation.

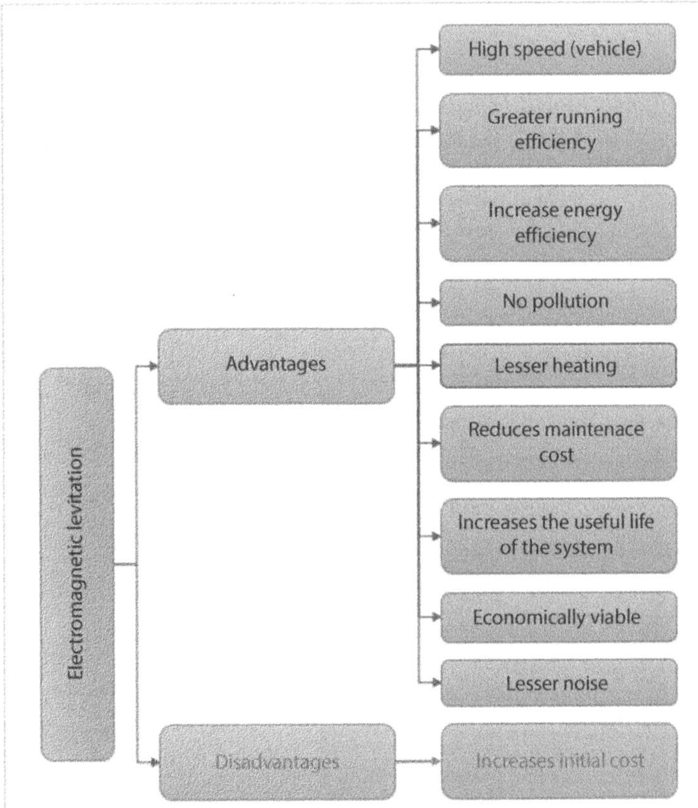

Figure 11.10 Advantages and disadvantages of levitation.

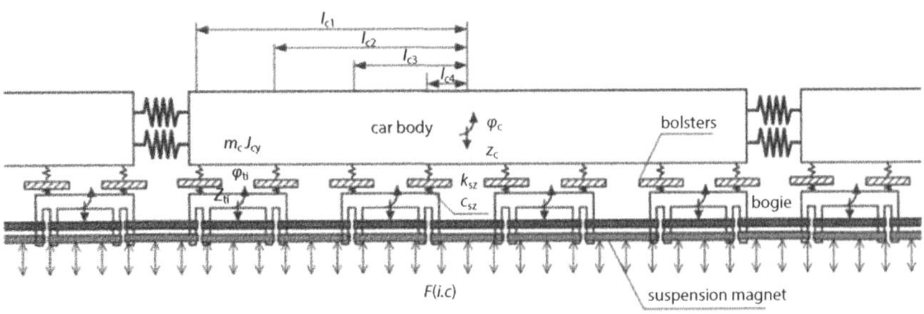

Figure 11.11 High speed train.

It is highly commendable that we are now on the brink of a transport revolution (Figure 11.13, Figure 11.15).

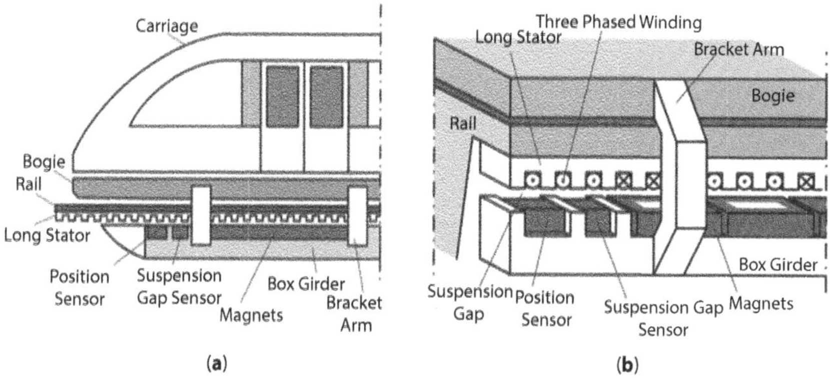

Figure 11.12 Model of high speed train (a) details of carriage and rail, (b) magnets and suspension system.

11.2 System Modelling and Fabrication

The prototype has been fabricated with spirit level accuracy, as shown in Figures 11.4 and 11.5.

The inductance profile inversely decreases when *air – gap* increases (Figure 11.23 (b)).

$$L(z) = \frac{N^2}{R_T} \tag{11.1}$$

Or,

$$L(z) = \frac{\mu_0 N^2 A R_T}{2z(t)} \tag{11.2}$$

where N is number of turns of the EM coil,

R_T is total reluctance path of the full magnetic circuit, z is *gap*, and A is the core area.

The electromagnetic attraction force developed between the flat steel plate and the EM coil is given by

$$F(i,z) = -\frac{d}{dz}\frac{1}{2}L(z)i(z)^2 = C\left(\frac{i}{z}\right)^2 \tag{11.3}$$

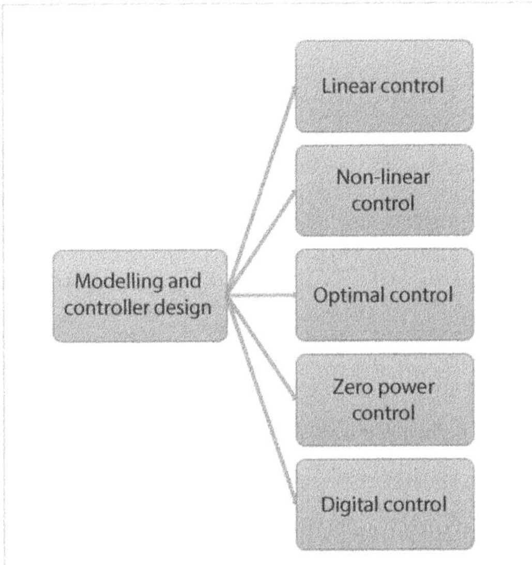

Figure 11.13 Different control strategies in levitation.

$$C = \frac{L_0 z_0}{2} = 1.89 \times 10^{-4} \tag{11.4}$$

FEM analysis (Figure 11.2) validates that the force value at the operating point (8 *mm*) is 1.45*N* (*mg*), which is the levitated force (weight of the plate) (Figure 11.14).

The electromagnetic coil has been designed with precise parameters (Table 11.1) and the desired specifications of the levitation system are shown in Table 11.2. The practical current vs. *air gap* is linear, as shown in (Figure 11.23 (a)). The equations of electro-mechanical system dynamic for the levitation system (Figure 11.1) are as follows:

$$m\frac{d^2z}{dt^2} = mg - F \tag{11.5}$$

$$Ri + L\frac{di}{dt} = v \tag{11.6}$$

From the mechanical dynamics, it is clearly shown that the system is highly non-linear, as represented by (11.5) and Figure 11.16). Thus, the

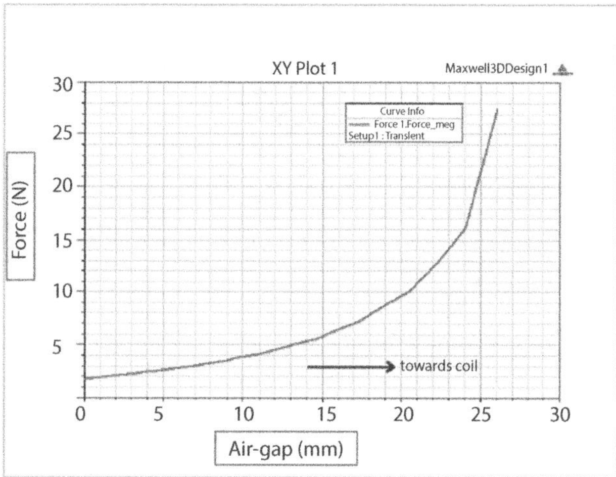

Figure 11.14 Force *air – gap* profile in FEM analysis.

Table 11.1 Parameters of magnetic coil.

Specifications	**Values**
Levitation Force (F)	1.45 N
Maximum Current Rating (Coil)	3 A
Number of Turns (EM Coil)	1500
Gauge of Copper Wire	22 SWG

Table 11.2 Design specifications.

System parameters	**Parameter values**
Mass (m) of Plate	148 gm
Position (z_0)	8 mm
EM Current (i_0)	0.6 A
Resistance (Coil)	11 Ω
Inductance (*Coil*)	0.06 H
Sensitivity of Feedback Sensor (B)	1000 V/m

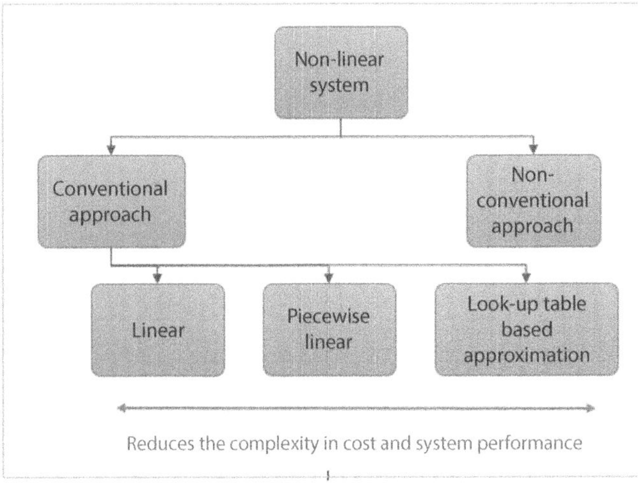

Figure 11.15 Different approach in modeling of levitation.

equations has been linearised at a suitable desired operating point and then taking Laplace Transformation the T.F. of the electrical dynamics $G(s)$ becomes

$$G(s)=\frac{\Delta I(s)}{\Delta V(s)}=\frac{1}{R+sL_0}=\frac{1}{11+0.078s} \tag{11.7}$$

The TF of the mechanical system for the perturbed/variational model, $G_1(s)$, of the plant is given by

$$G_1(s)=\frac{\Delta Z_0(s)}{\Delta I_0(s)}=-\frac{kz_0}{ms^2z_0-ki_0} \tag{11.8}$$

where $\Delta Z_0(s)$ implies small variations in *air → gap*. The force exponentially increases if *air gap* decreases.

Thus,

$$k=\frac{2Ci_0}{z_0^2} \tag{11.9}$$

Substituting the parameter in Table 11.2 in (11.4), we get

$$k = 4.13 \tag{11.10}$$

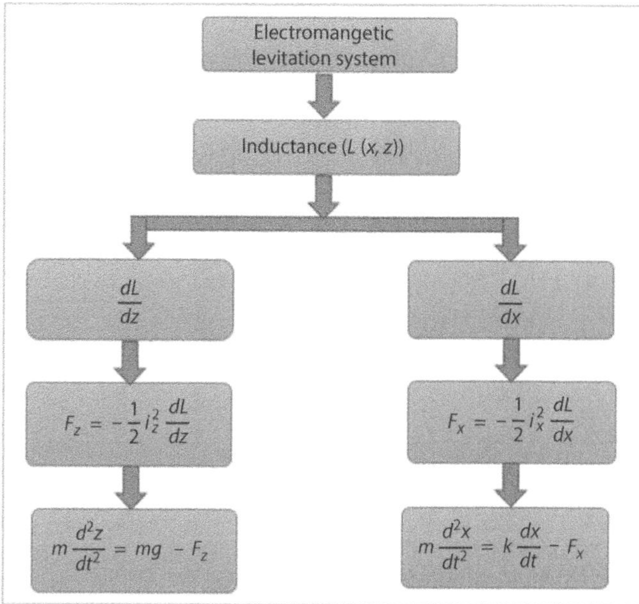

Figure 11.16 Modeling in *x* and *z* directions.

Substituting the system parameters (Table 11.2 in (11.8)) at the operating levitated point, the transfer function $G_2(s)$ becomes

$$G_2(s) = \frac{\Delta Z(s)}{\Delta I(s)} = \frac{28}{(s+49)(s-49)} \tag{11.11}$$

11.3 Feedback Sensing, Experimental Results, and Discussions

The study on stability for perturbations in the *x*-direction has been broadly studied in this work. Inductance $\frac{dL}{dx}$, restoring force (F_x) and current for small perturbations in the *x*-direction at different operating points (z) has been discussed broadly, which is shown in Figures 11.17, 11.19, 11.22. Velocity and acceleration plots at different operating air – gap are shown in Figures 11.18, 11.20, and 11.21. Figure 11.23 shows the inductance $\frac{dL}{dx}$, restoring force (Fx), and current vs. z_x– direction

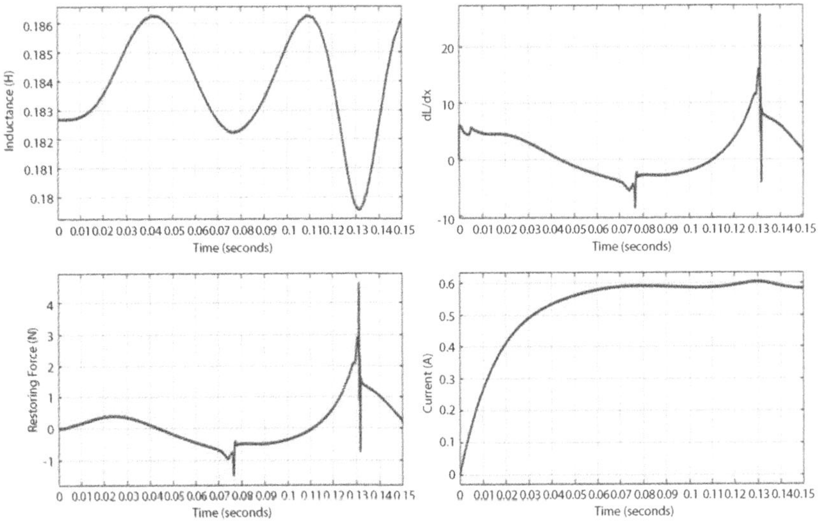

Figure 11.17 Different parameters vs. x-direction at $z = 2\ mm$.

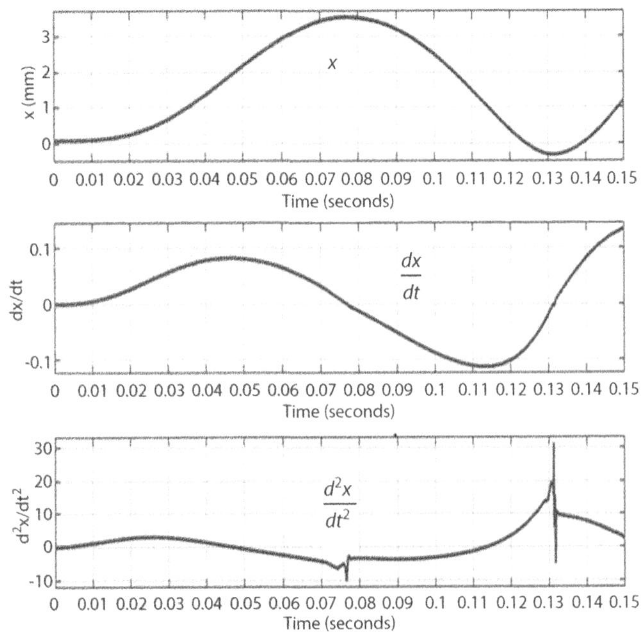

Figure 11.18 Velocity and acceleration plots at $z = 2\ mm$.

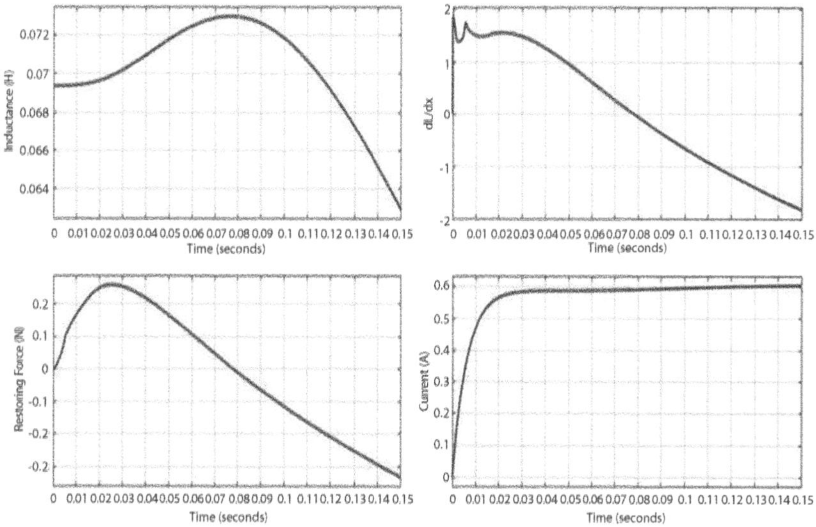

Figure 11.19 Different parameters vs. x-direction at $z = 7\ mm$.

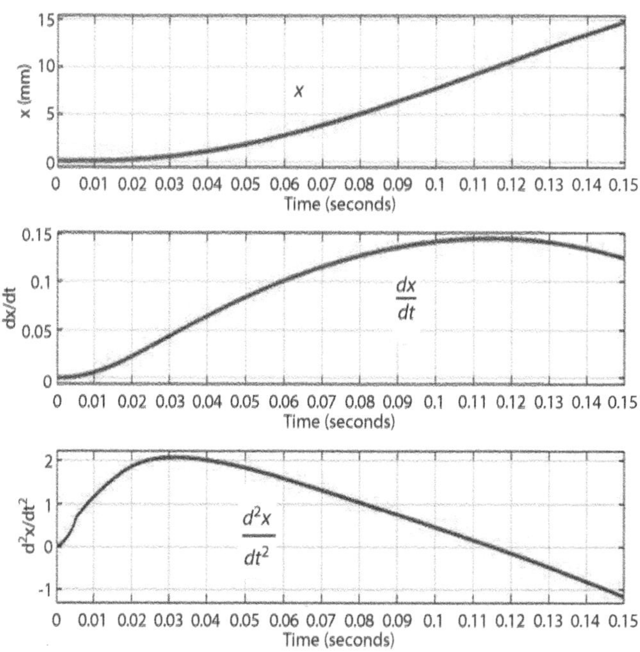

Figure 11.20 Velocity and acceleration plots at $z = 7\ mm$.

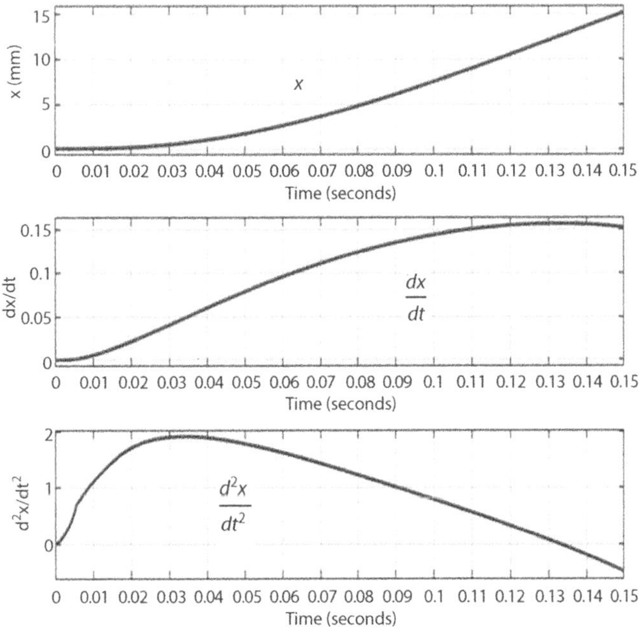

Figure 11.21 Velocity and acceleration plots at $z = 8\ mm$.

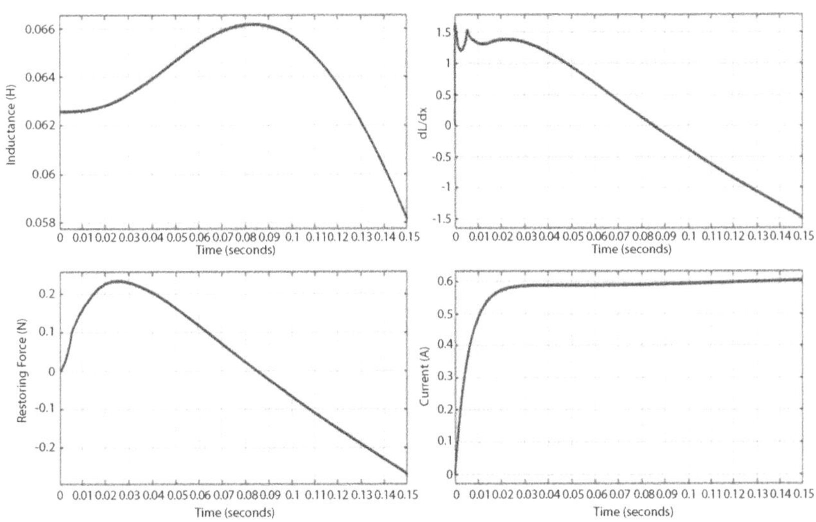

Figure 11.22 Different parameters vs. x-direction at $z = 8\ mm$.

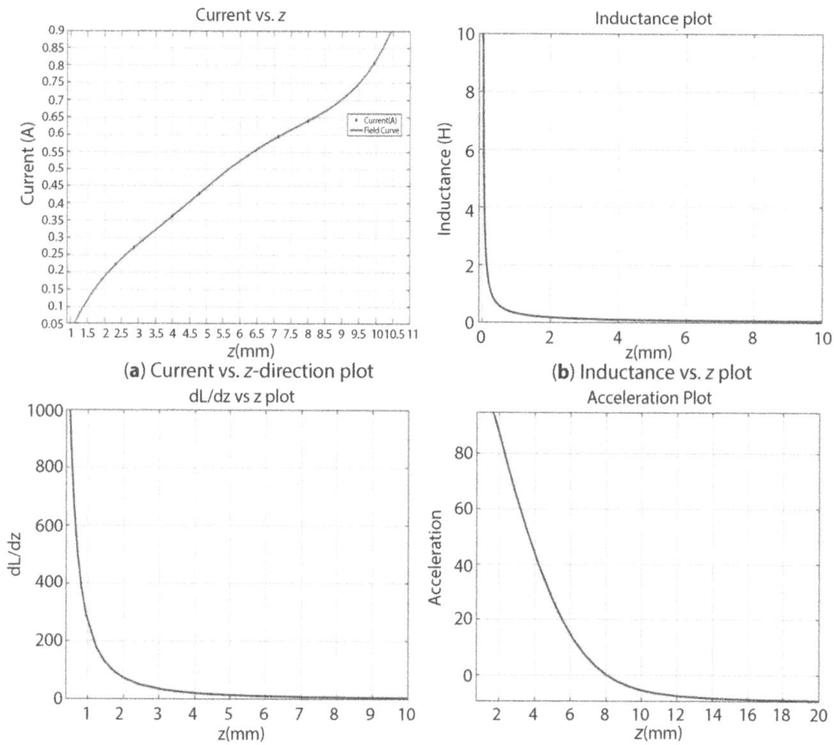

(**a**) Current vs. *z*-direction plot
(**b**) Inductance vs. *z* plot
(**c**) dL/dz vs. *z* plot
(**d**) Force vs. *z* plot at operating current

Figure 11.23 Different parameters vs. *z*-direction of flat plate levitation system.

plots. 2 *IR-LED* phototransistor pairs have been provided across the levitated object to restrict the movement in the x and y-directions, while a proximity sensor has been put in the z-direction. 2 pairs are for the x-axis, while 2 more pairs are for y-axis position detection. At the operating point, all the output signals from the sensors are being added and amplified. Then, the output signal is passed through a window comparator which makes the decision if it is at operating point or not. When the object is at the operating point, then sensors give a fixed $5V$ output, which is compared with the reference voltage and then controller works accordingly. The window comparator creates a stability zone around the operating point. The system has been linearised at an 8 *mm* operating gap. The control is more involved for an asymmetric shaped object than a spherical shaped object. A two loop controller (Figures 11.24 and 11.25) where the inner current control loop improves the system performance and the outer position control loop improve stability is used. Electrical dynamics are faster than the outer loop mechanical dynamics. Parameter uncertainty and robustness issues are also considered carefully. The gain margin (GM) of 60 *dB* and phase margin (PM) of 30° has been chosen for the overall system as per the standard classical control systems. Finally, a simple Darlington-pair driver circuit (Figure 11.26) has been designed for this system and an MJE13005 transistor provides the operating current of 0.6 *A* at a stable levitated position of 8*mm*. The transistor acts as a linear amplifier at the

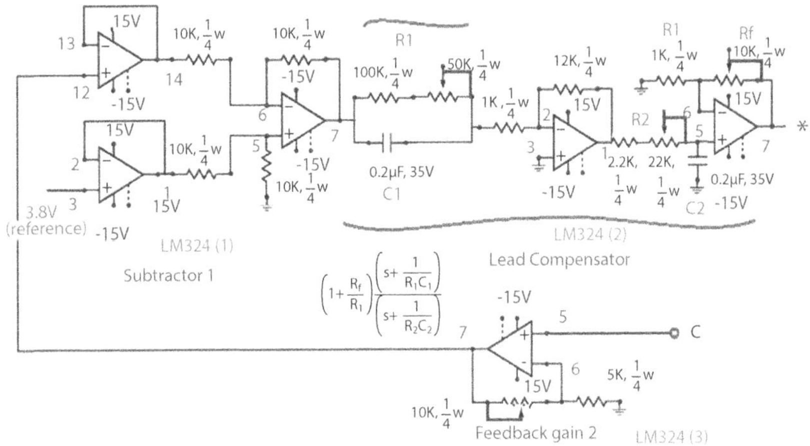

Figure 11.24 Circuit diagram of outer loop of controller.

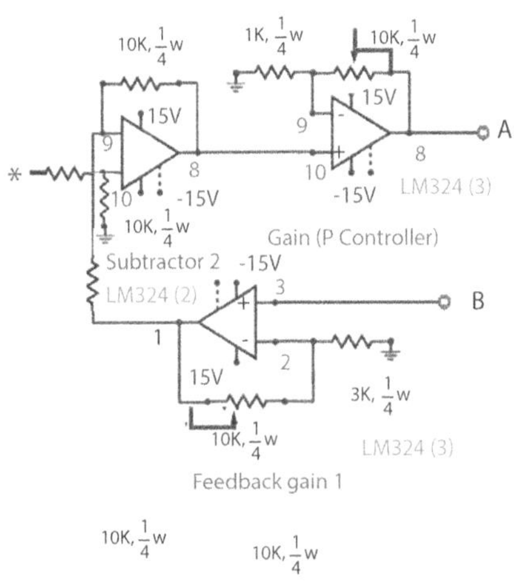

Figure 11.25 Circuit diagram of inner loop of controller.

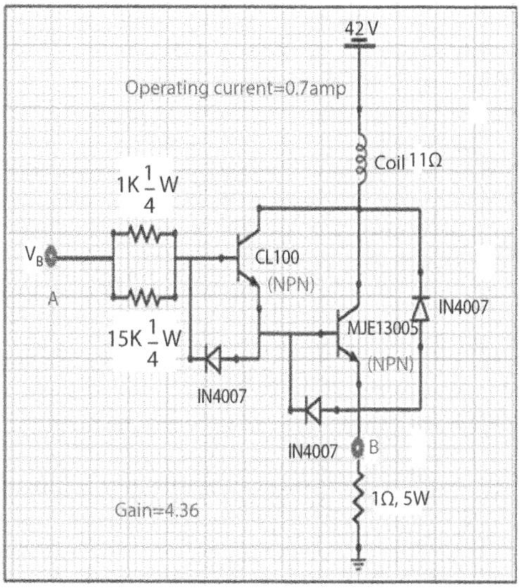

Figure 11.26 Driver circuit.

operating point. Sensitivity of the feedback sensors has been measured precisely and a weighted gain approach has been implemented to sense the feedback signal. Mechanical as well as electrical disturbances have been taken care of by suitable precautions in the set up. Parameters are evaluated by different methods to avoid parameter uncertainty. True mechanical zero and spirit level accuracy set up have been fabricated with smooth mechanical adjustments for the position of the plate.

11.4 Conclusions

All the parameters have been evaluated by three different methods and the plots match accurately, which lead to precise system modelling and controller design. Practical force vs. *air gap* has been shown to perfectly match with the FEM simulated plot (Figure 11.14). The practical force value is 1.45 *N*, where the simulated result is also 1.45 *N* at the operating point, which is the levitated force (weight of the plate). The practical plots are taken with very precise measurements with spirit level accuracy. The mechanical prototype has been fabricated in MSME Tool Room and Kolkata from a 3D AutoCAD drawing. True mechanical zero set up (Figure 11.3) has been fabricated. Sensors have been placed around the plate to create a stability zone. Performances of parameters due to mechanical disturbances in the

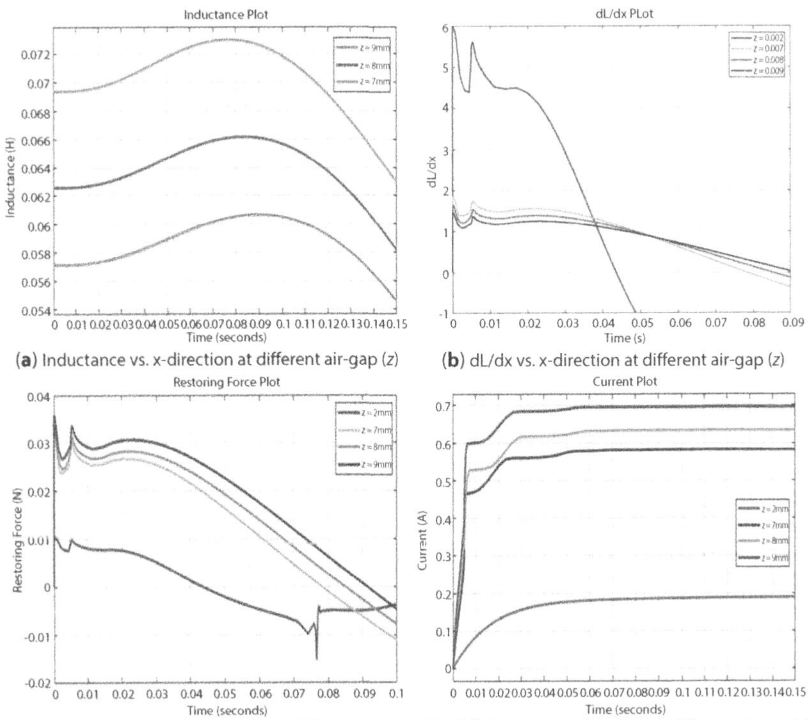

(**a**) Inductance vs. x-direction at different air-gap (*z*)

(**b**) dL/dx vs. x-direction at different air-gap (*z*)

(**c**) Restoring force (Fx) vs. x-direction at different air- gap (*z*)

(**d**) Current vs x-direction at different air-gap (*z*)

Figure 11.27 Comparative study of parameters at different operating air– gap.

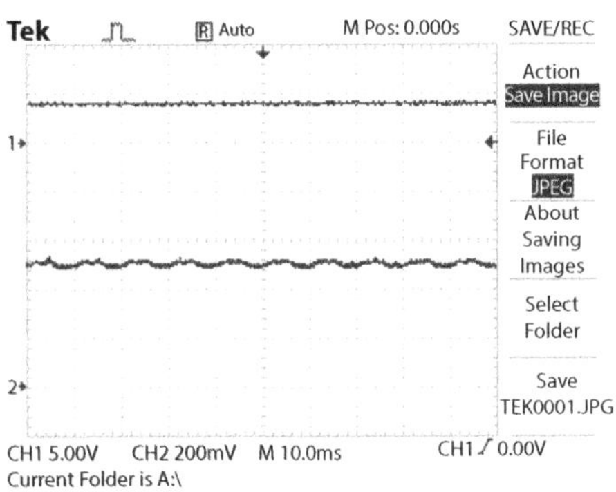

Figure 11.28 Operating current at stable levitated position.

x-direction have been highlighted in this work broadly. Velocity and acceleration plots at different operating points are highlighted (Figures 11.18, 11.20, and 11.21) in this work. Design of a weighted based controller and

Figure 11.29 Photograph of practical set up.

fully fabricated attraction type levitation prototype is developed where a flat steel plate has been successfully levitated. The controller has been implemented and simulated and its performance has been practically tested. Performance for transverse movement has been also studied in Figure 11.27. The experimental stable operating current is shown in Figure 11.28 and Figure 11.29 shows the fully lab developed practical prototype.

References

1. A. Bittar, JJ. Cruz, and RM. Sales, "A new approach to the levitation control of an electromagnetic suspension vehicle," *in Proc. of IEEE Int. Conf. on control applications*, pp. 263-270, 1998.
2. B. V. Jayawant, "Review lecture on electromagnetic suspension and levitation tech- niques," *in Proc. R. Soc. Lond.*, vol. 416, pp. 245-320, 1988.
3. J. Kundu, M. Sengupta, and A. Sengupta, "Design, modelling, fabrication and testing of an electromagnetic coil system for a levitation mechanism," *in Proc. Natl. Conf. NPEC*, Dec. 2013.
4. F. C. Moon *et al., Superconducting Levitation, Applications to Bearings and Magnetic Transportation,*" USA: John Wiley and Sons, 1994.
5. J. Pal, D. Prasad, and S. Banerjee, "A unifying approach for development of magnetically suspended vehicle using controlled DC electromagnet," *in Proc. Int. Conf. IEEE-ACE (EPIC)*, pp. 404-408, 2002.
6. D. Rogg, "General survey of the possible applications and development tendencies of magnetic levitation technology," *IEEE Trans. Magnetics*, vol. 20, pp. 1696-1701, Sep. 1984.

7. M. Sengupta and A. Sengupta, "Design, implementation and testing of an attraction type magnetic levitation system," *in Proc. Int. Conf. EAIT*, pp. 17-20, Dec. 2001.
8. P. K. Sinha *et al., Electromagnetic Suspension: Dynamics and Control.* London, U.K.: Peter Peregrinus, 1987.
9. T. H. Wong, "Design of a magnetic levitation control system," *IEEE Trans. Educ.*, vol. E-29, no. 4, pp. 196-200, Nov. 1986.
10. T. Ohtsuka, Y. Kyotani, "Superconducting levitated high speed ground transportation project in Japan", IEEE Transactions on Magnetics, vol. MAG-11, no. 2, Mar. 1975, pp. 608-614.
11. G. Ma, Z. Wang, K. Liu, H. Qian, C. Wang, "Potentials of an integrated levitation, guidance, and propulsion system by a superconducting transverse flux linear motor", IEEE Transactions on Industrial Electronics, Vol. 65, No. 9, Sep. 2018, pp. 7548-7557.
12. Jinhui L., Jie L., Danfeng Z., Peng C., Lianchun W., Peichang Y., "The active con- trol of maglev stationary self-excited vibration with a virtual energy harvester", IEEE Transactions on Industrial Electronics, Vol. 62, No. 5, May 2015, pp. 2942-2951.
13. M. Mofushita, T. Azukizawa, S. Kanda, N. Tamura, T. Yokoyama, "A new maglev system for magnetically levitated carrier system", IEEE Transactions on Vehicular Technology, Vol. 38, No. 4, Nov. 1989, pp. 230-236.
14. V. A. Oliveira, E. F. Costa, and J. B. Vargas, "Digital implementation of a magnetic suspension system for laboratory experiments," *IEEE Trans. Educ.*, vol. 42, pp. 315-319, Nov. 1999.
15. Yang, H. J., Lee, Y. S. and Kwon, O. K. "Development of force modeling equipment for magnetic levitation system", International Conference on Control and Automation and Systems, Korea, pp. 29-33, 2010.
16. F. Matsumura and S. Yamada, "A method to control the suspension system utilizing magnetic attractive force," *in Proc. Elect. Engg.*, vol. 94, pp. 50-54, 1974.
17. S. Moriyama, "AC magnetic suspension using a differential feedback power amplifier," *in Proc. Int. Conf. Elect. Engg.*, vol. 131, pp. 61-67, 2000.
18. O. Tsukamoto, K. Yasuda, and J. Z. Chen, "A new magnetic levitation system with AC magnets," *IEEE Trans. Magnetics*, vol. 24, pp. 1497-1502, Mar. 1988.
19. S. Yamamura, H. Yamaguchi, "Electromagnetic levitation system by means of salient-pole type magnets coupled with laminated slot less rails," *IEEE Trans. Vehicular Tech.*, vol. 39, pp. 83-88, Feb. 1990.
20. J. Chen and O. Tsukamoto, "Attractive type magnetic levitation system with 3 AC electro-magnets," *in Proc. IEEE Int. Conf. IECON*, vol. 1, pp. 778-782, Nov. 1991.
21. Y. S. Shiao, "Design and implementation of a controller for a magnetic levitation system," *in Proc. Natl. Sci. Counc.*, vol. 11, pp. 88-94, 2001.
22. M. Morishita and T. Azukizawa, "Zero power control method for electromagnetic levitation system," *IEE Proc.*, vol. 108, no. 45, pp. 447-451, 1988.

23. J. Jo, Y. Han, C. Lee , B. Kang, K. Kim, Y. Nam, S. Park, S. Lee, "Design and control of the miniature maglev using electromagnets and permanent magnets in magnetic levitation system", International Conference on Control, Automation and Systems Oct. 2010, pp. 27-30, KINTEX, Gyeonggi-do, Korea.
24. J. Jo, Y. Han, C. Lee, B. Kang, C. Kim, H. Cho, J. Lee, H. Han, B. Kim, D. Kim, "Zero-Power Control of Magnetic Levitation Vehicles with Permanent Magnets", International Conference on Control, Automation and Systems, Oct. 2010, 27-30, in KINTEX, Gyeonggi-do, Korea.
25. S. Banerjee, T. K. S. Kumar, J. Pal, and D. Prasad, "Controller design for large-gap control of electromagnetically levitated system by using an optimization technique," *IEEE Trans. Control Systems Technology*, vol. 16, no. 3, pp. 408-415, May. 2008.
26. J. Kundu, M. Sengupta "Design, Modelling, Fabrication of a Single-axis Attraction type Levitation Prototype with 3-degree Freedom Control," *in Proc. Natl. Conf. NPEC*, IIT Bombay, 2015.
27. J. Kundu, M. Sengupta, and A. Sengupta, "Design, modelling, fabrication and control of an attraction type levitation prototype," *in Proc. IEEE Intl. Conf.*, IIT Bombay, PEDES 2014.
28. K. Venkataratnam, T. K. Bhattacharya and M. Sengupta, "Theory, performance prediction and indirect sensing of rotor position in a switched reluctance motor under bulk saturation," *IEE Proc., EPA*, November, 1998.
29. Nicolsky R., Gorelov Y., Pereira A. S., David D. F. B., Santisteban A., Stephan R. M., Ripper A., De Andrade R. Jr., Gawalek W., Habisreuther T., Strasser T., "Super-conducting axial bearing for induction machines with active radial magnetic bearings", IEEE Transactions on Applied Superconductivity, Vol. 9, No. 2, Part 1, Jun 1999, pp. 964-967.
30. Nicolsky R., De Andrade R. Jr., Ripper A., David D. F. B., Santisteban J. A., Stephan R. M., Gawalek W., Habisreuther T., Strasser T., "Superconducting electromagnetic hybrid bearing using YBCO bulk blocks for passive axial levitation", Superconductor Science and Technology, Vol. 13, No. 6, Jun 2000, pp. 870-874.

12

Design of SSA Tuned Cascaded TI-TID Controller for Load Frequency Control of Multi-Source Power System with Electric Vehicle

Sandhya Kumari and Gauri Shankar*

Department of Electrical Engineering, Indian Institute of Technology (ISM), Dhanbad, Jharkhand, India

Abstract

The current work focuses on the usage of a cascade tilt-integral-tilt-integral-derivative (CTI-TID) controller for performance scrutiny of the load frequency control (LFC) of a two-area multi-source interlinked power system (MSIPS) consisting of thermal-hydro-gas generating units with nonlinearities. The structure of the CTI-TID controller is based on two methodologies: cascade control and fractional order calculus. The CTI-TID design uses TI and TID components representing slave and master controllers, respectively. The gain values of the designed CTI-TID controller are tuned by a salp swarm algorithm (SSA) for efficient operation. Firstly, the comparative performance study of SSA with that of other studied water cycle algorithms (WCA) and teaching learning based optimization (TLBO) techniques in tuning the proposed CTI-TID controller is carried out to confirm the competency of SSA. Secondly, dynamic responses obtained using an SSA tuned CTI-TID controller are compared to the responses found by the adopted SSA optimized cascade proportional-integral-tilt-derivative (CPI-TD) controller, proportional-integral-derivative (PID) controller, and other published works to determine its superiority in the MSIPS model. The simulation results obtained confirm the effectiveness of the proposed SSA tuned CTI-TID controller. Further, the effect of electric vehicles on the studied MSIPS under a CTI-TID controller is also examined. Lastly, to showcase the robustness of the proposed approach, sensitivity analysis is carried out.

**Corresponding author*: gaurishankar@iitism.ac.in

Vipin Chandra Pal, Suman Lata Tripathi, and Souvik Ganguli (eds.) Industrial Control Systems, (255–284) © 2024 Scrivener Publishing LLC

***Keywords*:** Cascade controller, electric vehicle, fractional order cascade controller, load frequency control, salp swarm algorithm

12.1 Introduction

The power system network is a complicated and massive system that connects several control areas via tie-lines to exchange power within them. The complexity of today's interlinked power system (IPS) has increased dramatically as a result of rising energy demands owing to the growing world's population. Furthermore, load demand is constantly changing and deviation in any area of IPS may results in transient deviations in its generation, frequency, and tie-line power. By controlling the MW power output of generators against load discrepancy, each area's frequency and scheduled tie-line power flow are kept within the stipulated limits through load frequency control (LFC) operation and a pertinent service executed by power system engineers [1–3]. The choice of controller is very important for operation and control of any industrial process including power systems. A variety of control strategies for LFC have been found in literature to preserve the frequency and tie-line power of IPS at a specified value under both normal and perturb situations [4]. In traditional control methods, integral (I) [5], proportional-integral (PI) [6], and dual mode PI [7] controller based strategies have been examined. The efficacy of a proportional-integral-derivative (PID) controller with that of I and PI controllers for LFC study are validated in [8, 9]. The comparative study of I, PI, integral-double-derivative, and PID controllers are reported in [10]. The presence of complexity such as nonlinearities owing to governor dead band (GDB), boiler dynamics (BD) and generation rate constraint (GRC), time delay, and load variations in IPSs degrades the performance of traditional controllers significantly. Some researchers have proposed cascade control approaches based on two feedback control loops to address these issues. Because of the large number of tuning knobs, cascade controllers have a good ability to outperform non-cascaded controllers [11, 12]. Das *et al.* [13] and Padhy *et al.* [14] have proven that a cascade PI-proportional-derivative (PI-PD) controller is more effective than a PID controller for LFC applications. Tripathy *et al.* [15] have suggested PD with a filter-PI controller for frequency stabilization in a two-area IPS consisting of diesel and distributed generation units. The performance investigation of PD-PID and PD-PID with filter controllers for LFC of multi-area IPSs are carried out in [16] and [17], respectively.

Researchers from all across the world are continuing to work on improving the LFC performance in complicated IPSs. This has led to the development of a new class of controllers, one of which is fractional order (FO) calculus-based controllers. Topno and Chanana [18] have designed a tilt-integral-derivative (TID) controller for frequency stabilization in a hydro-thermal IPS. A FOID controller was developed by Debbarma *et al.* for the frequency regulation of IPS [19]. Morsali *et al.* [20] have designed an improved particle swarm optimization (IPSO) tuned FOPID controller for LFC investigation of two-area IPS considering GDB and GRC. Effective LFC operation of two-area thermal IPS consisting of GDB using a FOPID controller is addressed in [21]. For the LFC examination of two- and three-area interlinked micro-grid systems, Latif *et al.* [22, 23] have developed a FOPID controller. In addition, a cascade arrangement of various FO controllers is used in the IPS to efficiently reduce any error in area frequency and tie-line power. In [24], a cascade FOPI-FOPD controller is suggested for LFC of a three-area IPSs incorporating renewable energy sources (RESs). Nayak and Shaw [25] verified the performance of a cascaded PD-FOPID controller over FOPID and PID counterparts in LFC. Recently, the effectiveness of PI-TID, FOPI-FOIDN, integral-derivative with filter (IDN)-FOPD, I-tilt-derivative (TD), and tilt-integral (TI)-TD controller based LFC of multi-area IPSs are tested in [26–29], and [30], respectively. Saha and Saikia in [31] provided the comparative performance examination of two-degree-of-freedom PIDN-FOI, PIDN-FOI, and PIDN controllers in a time domain for the examination of LFC of two-area thermal-gas IPSs comprising of distributed generations and electric vehicles (EVs). Today, EVs are highly recommended energy storage systems by researchers to enhance system stability. They are environmentally friendly and have a low charging cost. By releasing and absorbing extra energy as needed, they can be employed to reduce short-term frequency transients caused by changing load conditions or RES output [14, 29, 32–34]. In line with the above, the present authors designed cascade a TI-TID (CTI-TID) controller for the performance scrutiny of LFC of two-area multi-source IPS (MSIPS) with and without EVs. The gain parameters of controllers must be fine-tuned to operate reliably and satisfactorily. Hence, many meta-heuristic optimization strategies such as bacterial foraging [5], imperialist competitive algorithms [9], flower pollination algorithms [13], modified gray wolf optimization [14], spider monkey algorithms [15], bat algorithms [16], differential evolution [18], water cycle algorithms [23, 28], sine-cosine algorithms [24], etc. have been developed for LFC study. Authors in [35] have proposed a salp swarm algorithm (SSA) for solving distinct engineering issues. Recently, authors in [26] and [36] have proven the efficacy of SSA over other algorithms. Therefore, in the present work, SSA is utilized for optimal tuning of the proposed CTI-TID controller.

The following is an outline of the rest of the chapter: Sections 12.2 and 12.3 present the modelling of the analysed power systems and EVs, respectively. Section 12.4 discusses the adopted control approach. The SSA technique is briefly described in Section 12.5. In Section 12.6, the simulated results and their analysis are given, followed by Section 12.7, which concludes the current work.

12.2 Modelling of Studied MSIPS

In this study, the performance of an SSA tuned CTI-TID controller is examined using two-area MSIPS with identical thermal-hydro-gas generating units. The transfer function based linearized model of the studied two-area MSIPS with nonlinearities like GDB and GRC is displayed in Figure 12.1. The authors in [20, 37, 38] present detail information of each part of the studied MSIPS and their nominal parameter values. The nominal values of the studied system parameters are mentioned in the Appendix.

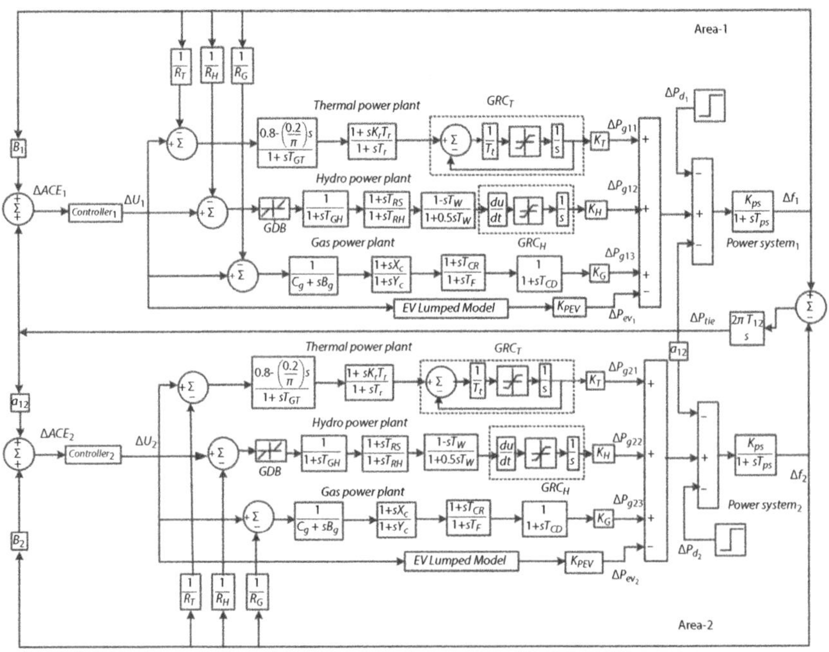

Figure 12.1 Configuration of studied two-area MSIPS.

For thermal units, the GRC limits are taken as 10% per minute, while for hydro units, the limit are 360% and 270% per minute for dropping and rising generation, respectively.

12.3 Modelling of EV

In this work, a lumped model of EVs as depicted in Figure 12.2 is considered. The flexibility of converters used in EVs to link in both directions has made them qualified as a reliable frequency support system. As a result, EVs can have an effect on frequency response in both charging and discharging modes. Figure 12.3 showcased the state transitions of EVs that contribute in LFC. In the driving state (i.e., uncontrollable state), EVs are disconnected from the power network. After the trip, the EVs switch from driving to charging mode (*i.e.*, plug-in) to recharge their battery and this time, they are unable to participate in frequency regulation. EVs react to control signals received from an electrical power system when they switch from charging to controllable mode (i.e., control-in). The distribution of control signals is depicted in Figure 12.4. The state-of-charge (SOC) of all controlled EVs is supposed to be synchronised in the studied EV model using the SOC synchronisation control method, which is detailed in [33]. Controllable EVs are those that have been charged up to 80% of their capacity. Additionally, EVs with an SOC greater than 80% are controlled in the 85 ± 5% range. The lower limit of 80% denotes that the EVs can be charged for a larger SOC before beginning the next journey, but the upper limit of 90% denotes that the battery lifetime decreases as the SOC approaches 100%. From Figure 12.2, ΔU_c, $\pm B_{kw}$, T_{ev}, E, and ΔP_{ev} are the LFC signal (*i.e.*

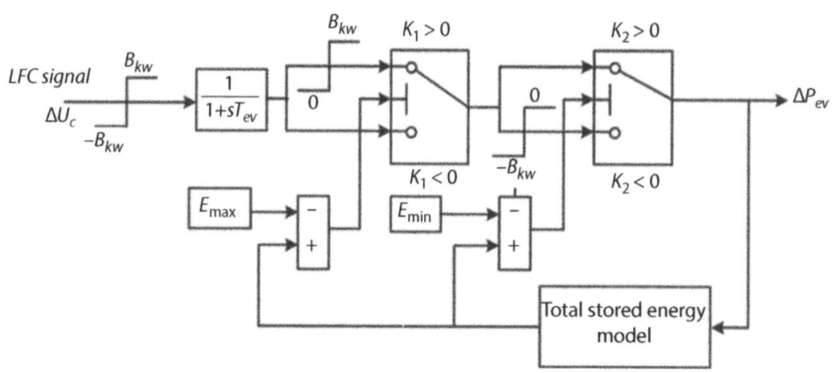

Figure 12.2 Configuration of lumped EVs model.

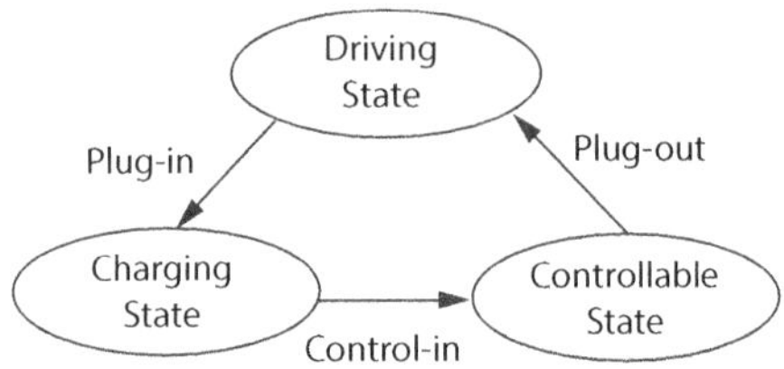

Figure 12.3 State and transitions of EVs.

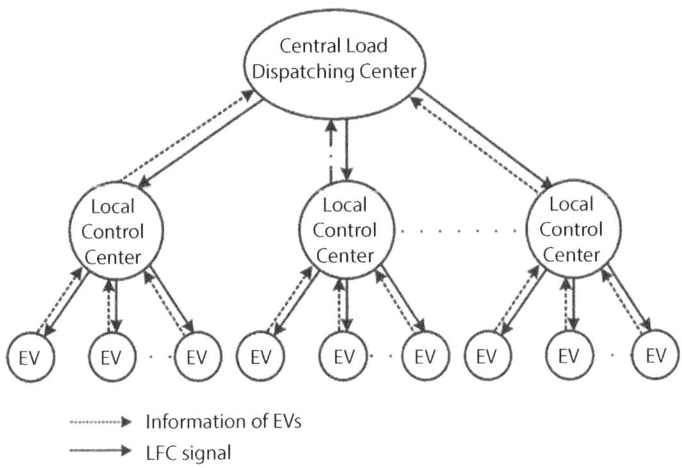

Figure 12.4 Configuration of control system of EVs.

ACE signal), limits of inverter capacity, time constant of EV, existing saved energy of the EV's battery, and the deviated power of EVs at the charging or discharging state, respectively. The total inverter capacity ($B_{kw}(t)$) of all controllable EVs is formulated by (12.1) [30].

$$B_{kw}(t) = N_{control}(t) \times B_{kw}^{*} \quad [kW] \tag{12.1}$$

where B_{kw}^{*} and $N_{control}(t)$ are the average capacity of the inverter and number of controllable EVs. $N_{control}(t)$ is determined by (12.2) [30].

$$N_{control}(t) = N_{initial}(t) + N_{control-in}(t) - N_{plug-out}(t) \tag{12.2}$$

Where $N_{initial}(t)$, $N_{control-in}(t)$ and $N_{plug-out}(t)$ are the number of initial, control-in, and plug-out EVs, respectively. The maximum (E_{max}) and minimum energy (E_{min}) of controllable EV batteries are stated by (12.3) and (12.4) [33].

$$E_{max}(t) = N_{control}(t) \times B^{*}_{kWh} \times 0.9 \quad [kWh] \tag{12.3}$$

$$E_{min}(t) = N_{control}(t) \times B^{*}_{kWh} \times 0.8 \quad [kWh] \tag{12.4}$$

where B^{*}_{kWh} is termed as average battery capacity. From Figure 12.2, K_1 and K_2 are defined by (12.5) [30].

$$K_1 = E_{control}(t) - E_{max}(t)$$
$$K_2 = E_{control}(t) - E_{min}(t) \tag{12.5}$$

where $E_{control}(t)$ is the entire stored energy of controllable EVs as depicted in Figure 12.5 and it is determined by (12.6) [33].

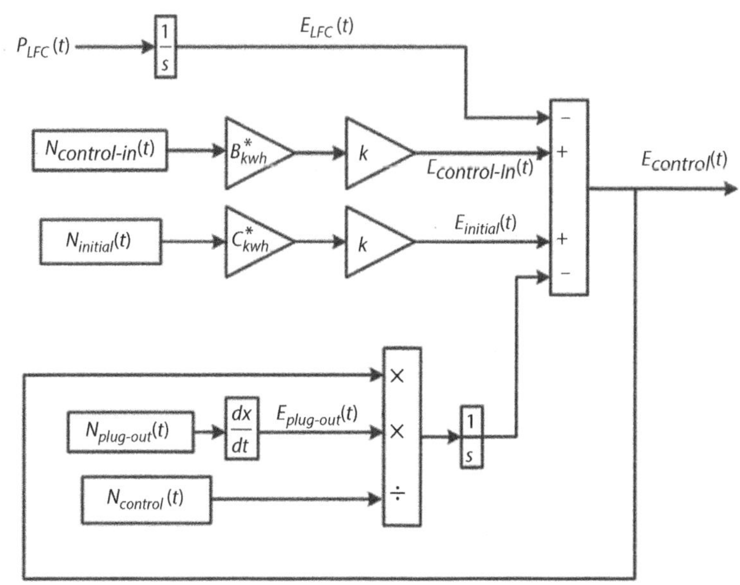

Figure 12.5 Stored energy model of one LC center.

$$E_{control}(t) = E_{initial}(t) + E_{control-in}(t) - E_{plug-out}(t) - E_{LFC}(t) \quad (12.6)$$

$E_{initial}(t)$, $E_{control-in}(t)$ and $E_{plug-out}(t)$ are computed by (12.7)-(12.9).

$$E_{initial}(t) = N_{initial}(t) \times B^{*}_{kWh} \times 0.85\,[kWh] \quad (12.7)$$

$$E_{control-in}(t) = N_{control-in}(t) \times B^{*}_{kWh} \times 0.85\,[kWh] \quad (12.8)$$

$$E_{plug-out}(t) = \int_0^t R_{plug-out}(t') \times \frac{E_{control}(t')}{N_{control}(t')} dt' \;\; [kWh] \quad (12.9)$$

where $R_{plug-out} = \frac{d}{dt}(N_{plug-out})$. The LFC signal energy is found by (12.10)

$$E_{LFC}(t) = \int_0^t P_{LFC}(\tau) d\tau \quad (12.10)$$

where P_{LFC} is the power of local control (LC) center. An EV responds to an LFC signal when $E_{control}(t)$ lies between E_{max} and E_{min}.

12.4 Adopted Control Approach

12.4.1 PID Controller

The configuration of the PID controller is described in Figure 12.6 and its transfer function is presented in (12.11).

$$T_{PID}(s) = k_p + \frac{k_i}{s} + k_d \frac{sN}{s+N} \quad (12.11)$$

where k_p, k_i, k_d, and N are the gains/filter coefficient of the PID controller.

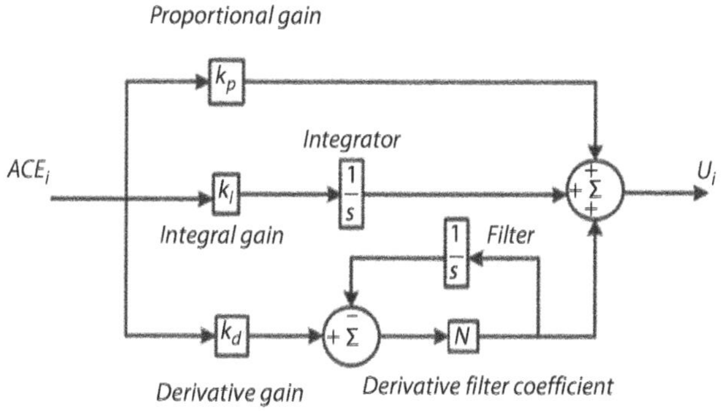

Figure 12.6 Configuration of PID controller.

12.4.2 Cascade Controller

The cascade control methodology is mostly utilised for quick disturbance rejection before it reaches the rest of the plant/process. The block representation of a cascade controller with plant/process is showcased in Figure 12.7 where $T_{p_1}(s), T_{p_2}(s), T_{c_1}(s)$ and $T_{c_2}(s)$ are the transfer functions of primary loop plant, secondary loop plant, master controller, and slave controller, respectively. The closed-loop transfer function of Figure 12.7 is stated by (12.12).

$$T_{cc}(s) = \frac{T_{p_2}(s)}{1 + T_{p_1}(s)T_{c_2}(s) + T_{p_1}(s)T_{p_2}(s)T_{c_1}(s)T_{c_2}(s)} D(s) + \frac{T_{p_1}(s)T_{p_2}(s)T_{c_1}(s)T_{c_2}(s)}{1 + T_{p_1}(s)T_{c_2}(s) + T_{p_1}(s)T_{p_2}(s)T_{c_1}(s)T_{c_2}(s)} R(s) \quad (12.12)$$

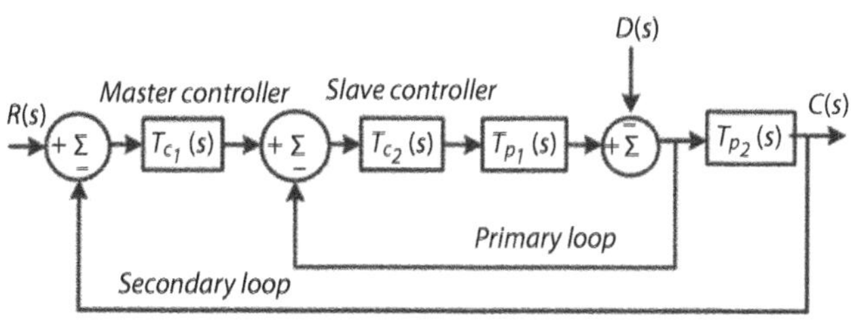

Figure 12.7 Configuration of cascade controller.

where $D(s)$ refers to load fluctuation and $R(s)$ indicates reference input to the process, respectively.

12.4.3 CPI-TD Controller

The configuration of CPI-TD controller as demonstrated in Figure 12.8 is based on both cascade and fractional-order control theory. $T_{c_1}(s)$ and $T_{c_2}(s)$ of the CPI-TD controller are presented by (12.13) and (12.14), respectively.

$$T_{c_1}(s) = k_p + \frac{k_i}{s} \tag{12.13}$$

$$T_{c_2}(s) = k_t s^{-1/n} + k_d \frac{sN}{s+N} \tag{12.14}$$

where k_t is the tilt gain of CPI-TD controller and the value of tilt parameter (*i.e., n*) lies between 2 and 3.

12.4.4 Design of CTI-TID Controller

Like the CPI-TD controller, the design of CTI-TID controller is also based on both the cascade and fractional-order control theory. The CTI-TID

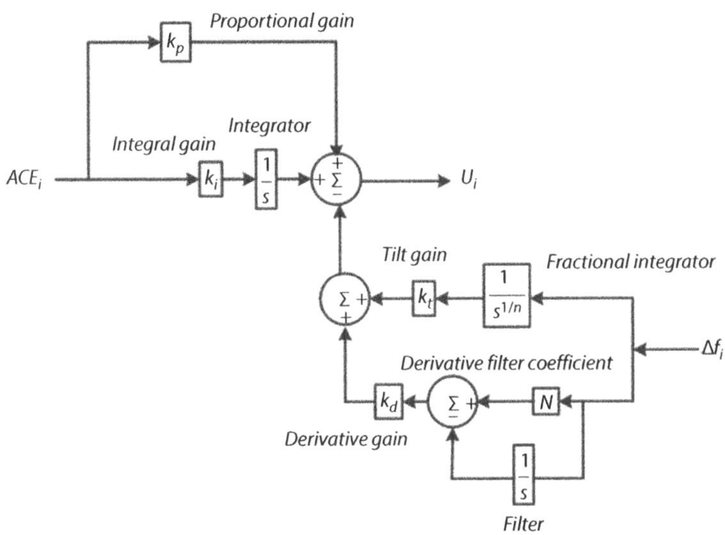

Figure 12.8 Configuration of CPI-TD controller.

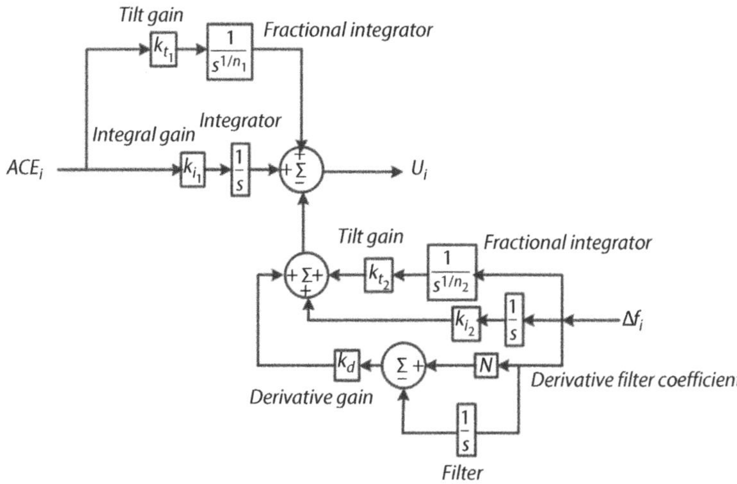

Figure 12.9 Configuration of CTI-TID controller.

controller is configured with proportional gain (k_p) (as in CPI-PID controller) substituted with the tilting gain parameter $\left(k_t s^{-1/n}\right)$ as shown in Figure 12.9 and $T_{c_1}(s)$ and $T_{c_2}(s)$ of CTI-TID controller are demonstrated by (12.15) and (12.16), respectively.

$$T_{c_1}(s) = k_{t_1} s^{-1/n} + \frac{k_{i_1}}{s} \tag{12.15}$$

$$T_{c_2}(s) = k_{t_2} s^{-1/n} + \frac{k_{i_2}}{s} + k_d \frac{sN}{s+N} \tag{12.16}$$

where k_{i_1} and k_{t_1} are named as integral and tilt gains of the master controller of the CTI-TID controller. Similarly, k_{i_2} and k_{t_2} indicate the integral and tilt gains of the slave controller of the CTI-TID controller.

12.4.5 Formulated Fitness Function and Optimization Constraint

The appropriate search of fitness function (or figure of demerit (FOD)) is important in getting optimal controller parameters to alleviate tie-line power and frequency errors. The fitness functions are considered the same as those used in [20, 38] for a better comparative study. The integral of time square error (ITSE) is used as FOD in this study and it is presented by (12.17).

$$FOD = \int_{0}^{T_{sim}} t\left(\Delta f_1^2 + \Delta f_2^2 + \Delta P_{tie}^2\right)dt \tag{12.17}$$

where T_{sim}, Δf_1, Δf_2, and ΔP_{tie} are the simulation time, area-1 frequency error, area-2 frequency error, and changes in tie-line power flow, respectively. FOD is subjected to constraint as stated in (12.18).

$$\left.\begin{array}{l}
k_{p,j}^{min} \le k_{p,j} \le k_{p,j}^{max} \\
k_{i,j}^{min} \le k_{i,j} \le k_{i,j}^{max} \\
k_{d,j}^{min} \le k_{d,j} \le k_{d,j}^{max} \\
k_{t,j}^{min} \le k_{t,j} \le k_{t,k}^{max} \\
k_{i_1,j}^{min} \le k_{i_1,j} \le k_{i_1,j}^{max} \\
k_{i_2,j}^{min} \le k_{i_2,j} \le k_{i_2,j}^{max} \\
k_{t_1,j}^{min} \le k_{t_1,j} \le k_{t_1,j}^{max} \\
k_{t_2,j}^{min} \le k_{t_2,j} \le k_{t_2,j}^{max} \\
n_{1,j}^{min} \le n_{1,j} \le n_{1,j}^{max} \\
n_{2,j}^{min} \le n_{2,j} \le n_{2,j}^{max}
\end{array}\right\} \tag{12.18}$$

where $k_{p,j}, k_{i,j}, k_{i_1,j}, k_{t_1,j}$ and $n_{1,j}$ are the gains of the master controller of jth-area. $k_{t_2,j}, k_{i_2,j}, k_{d,j}$, and $n_{2,j}$, are gains of the slave controller of jth-area. The minimum and maximum bound of the gain values of the studied PID/CPI-TD/proposed CTI-TID controllers are indicated by *min* and *max*, respectively. Other performance indicators like integral absolute error (IAE), integral time absolute error (ITAE), and integral square error (ISE) are determined after the termination of the main program to prove the ability of the proposed SSA tuned CTI-TID controller. The IAE, ITAE, and ISE are given by (12.19)-(12.21).

$$IAE = \int_{0}^{T_{sim}} (|\Delta f_1| + |\Delta f_2| + |\Delta P_{tie}|)dt \tag{12.19}$$

$$ITAE = \int_{0}^{T_{sim}} t(|\Delta f_1| + |\Delta f_2| + |\Delta P_{tie}|)dt \tag{12.20}$$

$$ISE = \int_{0}^{T_{sim}} \left(\Delta f_1^2 + \Delta f_2^2 + \Delta P_{tie}^2\right)dt \tag{12.21}$$

12.5 Description of SSA

SSA is a recently developed meta-heuristic optimization method that is motivated by the swarming existence of salps in the deep ocean [35]. It has a translucent barrel-shaped shell, similar to jellyfish. The salp chain is a swarm of salps that form within the ocean. While modelling salp chains, the generation of salp is divided into two sets: the leader and the follower. The leader of these salps leads the swarm (i.e., presents in front location), while the followers follow one another. The current leader's position is mentioned in (12.22) [36].

$$Z_k^1 = \begin{cases} F_k + C_1((UL_k - LL_k)C_2 + LL_k)\ if\ C_3 \geq 0.50 \\ F_k - C_1((UL_k - LL_k)C_2 + LL_k)\ if\ C_3 < 0.50 \end{cases} \tag{12.22}$$

where Z_k^1, F_k, LL_k and UL_k, are the first salp (leader) location, position of the food source, and the lower and upper limits of kth dimension, respectively. C_1 is a crucial parameter in SSA since it ensures that the exploration and exploitation phases are balanced. $C_{1,}$ C_2, and C_3 are determined by (12.23) [35].

$$\left.\begin{aligned} C_1 &= 2\exp^{-\left(\frac{4P}{M}\right)^2} \\ C_2, C_3 &\in [0,1] \end{aligned}\right\} \tag{12.23}$$

where P is the present generation number and M is the maximum generation number, respectively. The follower's position is now updated by (12.24) [35].

$$Z_k^m = \frac{1}{2}\left(\frac{V_{final}}{V_o}\right)T^2 + V_o T \tag{12.24}$$

where $m \geq 2$ and Z_k^m indicate the location of the mth follower salp in the kth dimension. $V_{final} = \frac{Z - Z_0}{T}$, T and V_o are the final velocity, time, and initial velocity, respectively. Time is the iteration in this optimization method and the difference within two consecutive generations is one. Therefore, (12.24) can be rewritten as (12.25), where $V_o = 0$.

$$Z_k^i = \frac{1}{2}\left(Z_k^i + Z_k^{i-1}\right) \tag{12.25}$$

The salp chain can be simulated utilizing (12.22) and (12.25).

12.6 Simulation Results and Analysis

In this work, three scenarios are scrutinized to validate the performance of the designed SSA tuned CTI-TID controller. The proposed CTI-TID and CPI-TD controllers are modelled using an n-integer toolbox [39]. The test model as displayed in Figure 12.1 is simulated in MATLAB/SIMULINK[R] and the program of the SSA is developed in *.m-file*. The SSA parameters are obtained after several trail runs and these are: population size=50 and number of iteration=100. For the purpose of proving the ability of the proposed approach, the work in this section is divided into the following categories.

Scenario 1: Performance Investigation of Studied Two-area MSIPS Model
Scenario 2: Performance Investigation of Studied Two-area MSIPS Model with EVs
Scenario 3: Sensitivity Analysis

Table 12.1 Optimized gain values of adopted algorithm of Scenario 1.

Parameters	TLBO		WCA		SSA	
Parameters Values	Area 1	Area 2	Area 1	Area 2	Area 1	Area 2
	$k_{t1} = 0.5525$	$k_{t1} = 0.8175$	$k_{t1} = 0.1616$	$k_{t1} = 1.4954$	$k_{t1} = 2.0000$	$k_{t1} = 1.8566$
	$n_1 = 2.1797$	$n_1 = 2.5618$	$n_1 = 2.1567$	$n_1 = 2.0862$	$n_1 = 3.0000$	$n_1 = 2.0076$
	$k_{i1} = 0.2810$	$k_i = 0.6315$	$k_{i1} = 0.5629$	$k_i = 0.1136$	$k_{i1} = 1.0211$	$k_i = 1.7578$
	$k_{t2} = 1.4102$	$k_{t2} = 1.1297$	$k_{t2} = 2.0000$	$k_i = 0.9873$	$k_{t2} = 1.9990$	$k_{t2} = 1.7448$
	$k_{i2} = 0.1971$	$k_{i2} = 1.4091$	$k_{i2} = 0.4183$	$k_{i2} = 0.1954$	$k_{i2} = 0.0669$	$k_{i2} = 0.0890$
	$n_2 = 2.4178$	$n_2 = 2.8096$	$n_2 = 2.4858$	$n_2 = 2.5346$	$n_2 = 2.9795$	$n_2 = 2.3676$
	$k_d = 1.4199$	$k_d = 1.4788$	$k_d = 1.4878$	$k_d = 1.4292$	$k_d = 1.6050$	$k_d = 1.6449$

Table 12.2 Transient response and performance indices for distinct algorithms related to two-are MSIPS of Scenario 1.

	Transient response									**Performance indices**			
	Us (p.u.)			*Os (p.u.)*			*Ts (s)*						
Studied algorithms	Δf_1 (×10^{-2})	Δf_2 (×10^{-2})	ΔP_{tie} (×10^{-2})	Δf_1 (×10^{-2})	Δf_2 (×10^{-2})	ΔP_{tie} (×10^{-2})	Δf_1	Δf_2	ΔP_{tie}	**FOD** (×10^{-2})	**ISE** (×10^{-4})	**ITAE**	**IAE** (×10^{-2})
TLBO	2.94	1.46	0.48	0.55	0.39	1.20	13.26	16.32	26.84	0.58	9.36	0.82	9.70
WCA	2.92	1.45	0.47	0.62	0.40	1.31	13.13	14.55	27.12	0.52	8.42	0.75	8.68
SSA	**2.87**	**1.34**	**0.44**	**0.65**	**0.45**	**7.43**	**13.84**	**19.45**	**23.15**	**0.43**	**7.10**	**0.66**	**7.72**

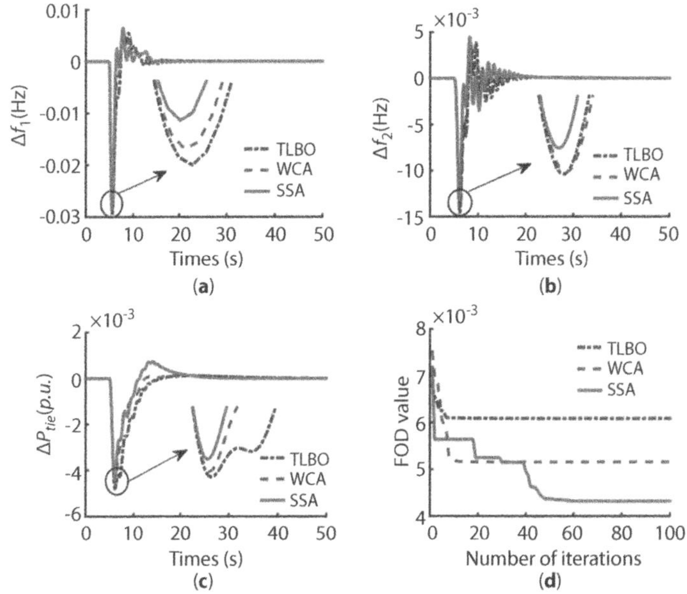

Figure 12.10 Comparative transient response profiles of two-area MSIPS obtained using studied algorithms of Scenario 1 (a) Δf_1 (b) Δf_2 (c) ΔP_{tie} (d) FOD.

12.6.1 Scenario 1: Performance Investigation of Studied Two-Area MSIPS Model

In the present section study, two types of investigation are carried out. Firstly, tuning efficiency of SSA is proved with respect to adopted TLBO and WCA. The details of WCA and TLBO are found in [29, 37]. To demonstrate the efficacy of the SSA over TLBO and WCA, a two-area MSIPS model is used to examine the performance of LFC. Testing is done with the CTI-TID controller only. Similar to [20, 38], a step load perturbation (SLP) of 0.01 p.u. at t =5 second (s) and FOD (as mentioned in (12.17)) as an objective function is considered for comparative analysis. The best optimized gain values of SSA, WCA, and TLBO tuned CTI-TID controllers is noted in Table 12.1. The comparative transient responses of Δf_1, Δf_2, and ΔP_{tie} in terms of undershoot (U_s), overshoot (O_s), and settling time (T_s) are tabulated in Table 12.2 and their comparative response profiles are shown in Figures 12.10 (a)-(c), respectively. The FOD values obtained under TLBO, WCA, and SSA tuned TI-TID controllers are compared in Figure 12.10 (d). The comparative values of FOD and others performance

Table 12.3 Optimized gain values of adopted controller of Scenario 1.

Parameters	I*		PID*		PID		CPI-TD	
Parameters Values	Area 1	Area 2	Area 1	Area 2	Area 1	Area 2	Area 1	Area 2
	$k_i = 0.0851$	$k_i = 0.0457$	$k_p = 0.8576$	$k_p = 0.8576$	$k_p = 1.8810$	$k_i = 0.0100$	$k_p = 1.9986$	$k_p = 1.6145$
			$k_i = 0.3189$	$k_i = 0.3057$	$k_i = 1.9950$	$k_i = 0.0020$	$k_i = 1.1809$	$k_i = 1.4755$
			$k_d = 0.6836$	$k_d = 0.5059$	$k_d = 1.9508$	$k_d = 1.9972$	$k_{t2} = 1.8982$	$k_{t2} = 0.6174$
							$n_2 = 2.0004$	$n_2 = 2.9133$
							$k_d = 1.5165$	$k_d = 1.4908$

'*' means taken from work of [20, 38].

Table 12.4 Optimized gain values of proposed CTI-TID controller with and without EVs of Scenario 1.

Parameters	**CTI-TID (without EVs)**		**CTI-TID (with EVs)**	
Parameters Values	Area 1	Area 2	Area 1	Area 2
	$k_{t1} = 2.0000$	$k_{t1} = 1.8566$	$k_{t1} = 1.9970$	$k_{t1} = 1.9433$
	$n_1 = 3.0000$	$n_1 = 2.0076$	$n_1 = 2.7604$	$n_1 = 2.2119$
	$k_{i1} = 1.0211$	$k_i = 1.7578$	$k_i = 0.4690$	$k_i = 0.7729$
	$k_{t2} = 1.9990$	$k_{t2} = 1.7448$	$k_{t2} = 2.0000$	$k_{t2} = 2.0000$
	$k_{i2} = 0.0669$	$k_{i2} = 0.0890$	$k_{i2} = 0.9408$	$k_{i2} = 0.2071$
	$n_2 = 2.9795$	$n_2 = 2.3676$	$n_2 = 2.9957$	$n_2 = 2.5915$
	$k_d = 1.6050$	$k_d = 1.6449$	$k_d = 1.4256$	$k_d = 1.7629$

indices are tabulated in the 11th to 14th columns of Table 12.2. Table 12.2 and Figure 12.10 illustrate that SSA is superior to the other algorithms. As a result, SSA is used only for further investigation.

Secondly, the supremacy of the CTI-TID controller is tested over the other studied controllers. For this, the same studied model is considered. The best optimized gain values of adopted PID, CPI-TD, and CTI-TID controllers along with the optimized gain values of I and PID controllers as reported in [20, 38] are noted in Tables 12.3-12.4. Figures 12.11 (a)-(c) portray the comparative response profiles of Δf_1, Δf_2, and ΔP_{tie} obtained using SSA tuned adopted PID, CPI-TD, and proposed CTI-TID controllers and other IPSO tuned conventional I and PID controllers, as stated in [20, 38]. The comparative profile of FOD obtained using adopted controllers is showcased in Figure. 12.11 (d). Table 12.5 presents the comparative dynamic responses of Δf_1, Δf_2, and ΔP_{tie} in terms of U_s, O_s, and T_s. The comparative values of FOD and other performance indices are noted in the 11th to 14th columns of Table 12.5. From Table 12.5, the FOD value found under an SSA tuned CTI-TID controller is 0.0024, which is 93.08%, 83.27%, 75.84%, and 8.51% better than IPSO tuned I, IPSO tuned PID, SSA tuned PID, and SSA tuned CPI-TD controllers, respectively. It is confirmed from Table 12.5 and Figure 12.11 that an SSA tuned CTI-TID controller provides better performance than others.

Table 12.5 Transient response and performance indices related to two-area MSIPS of Scenario 1.

Studied controllers	**Transient response**									**Performance indices**			
	$U_s(p.u.)$			$O_s(p.u.)$			T_s**(s)**						
	Δf_1 ($\times 10^{-2}$)	Δf_2 ($\times 10^{-2}$)	ΔP_{tie} ($\times 10^{-2}$)	Δf_1 ($\times 10^{-2}$)	Δf_2 ($\times 10^{-2}$)	ΔP_{tie} ($\times 10^{-2}$)	Δf_1	Δf_2	ΔP_{tie}	**FOD** ($\times 10^{-2}$)	**ISE** ($\times 10^{-4}$)	**ITAE**	**IAE** ($\times 10^{-2}$)
I*	4.51	5.82	0.88	1.09	1.09	4.20	48.48	50.00	49.78	6.21	-	-	-
PID*	3.84	4.37	0.72	1.95	1.79	15.0	28.03	26.81	40.78	2.57	-	-	-
PID	3.50	3.10	0.56	1.23	1.16	21.0	14.45	18.64	18.08	1.78	24.0	1.65	17.7
PI-TD	2.92	1.43	0.46	0.41	0.31	6.44	13.94	19.47	23.51	0.47	7.72	0.69	8.11
TI-TID (without EV)	2.87	1.34	0.44	0.65	0.45	7.43	13.84	19.45	23.15	0.43	7.10	0.66	7.72
TI_TID (with EVs)	**2.40**	**0.74**	**0.36**	**0.56**	**0.12**	**2.94**	**12.54**	**17.45**	**24.52**	**0.24**	**4.03**	**0.43**	**5.18**

'-' means not applicable.

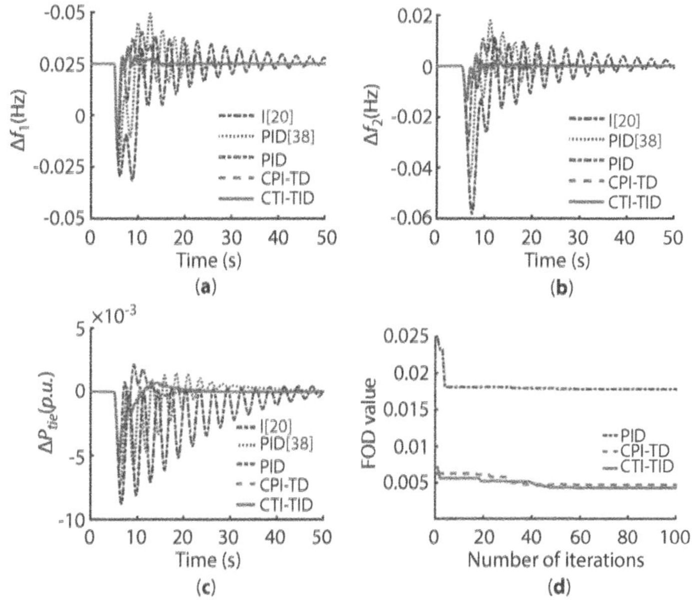

Figure 12.11 Comparative transient response profiles of two-area MSIPS obtained using studied controllers of scenario 1 (a) Δf_1 (b) Δf_2 (c) ΔP_{tie} (d) FOD.

12.6.2 Scenario 2: Performance Investigation of Studied Two-Area MSIPS Model with EVs

The work of Scenario 1 is further extended by inclusion of EVs in both areas of the studied two-area MSIPS model. In this scenario, the impact of EVs on the studied two-area MSIPS in the presence of a CTI-TID controller is examined. The participation factor of thermal, hydro, gas units, and EVs are taken as 0.4539, 0.3261, 0.1304, and 0.09, respectively. The nominal parameters of EVs are mentioned in the Appendix section. The same SLP and FOD function as considered in Scenario 1 are used for the present analysis. The optimized gain values of the proposed SSA tuned CTI-TID controller in the presence of EVs are given in the 3rd column of Table 12.4. The response profiles of Δf_1, Δf_2, and ΔP_{tie} and FOD obtained using the proposed SSA tuned CTI-TID controller with and without EVs are compared in Figure 12.12. It is seen from the 9th row of Table 12.5 that U_s, O_s and T_s of Δf_1, Δf_2, and ΔP_{tie} are improved in the presence of EVs. From the 9th row of Table 12.5, the improvement seen due to the presence of EVs in (i) FOD is 44.19%, (ii) ISE is 43.24%, (iii) ITAE is 34.85%, and (iv) IAE is 32.90%. It is evident from Figure 12.12 and Table 12.5 that the transient responses of studied MSIPS are enhanced in the presence of EVs.

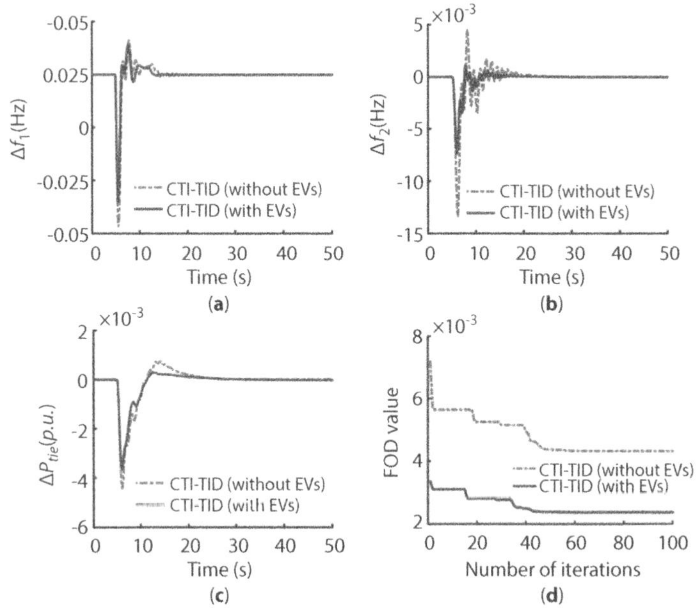

Figure 12.12 Comparative transient response profiles using proposed controller with and without EVs of Scenario 2 (a) Δf_1 (b) Δf_2 (c) ΔP_{tie} (d) FOD.

12.6.3 Scenario 3: Sensitivity Analysis

The robustness of the proposed SSA optimized CTI-TID controller against large changes in loading situations and parameters of the analysed two-area MSIPS with EVs is validated using sensitivity analysis.

The considered parameters are loading condition, synchronising coefficient (T_{12}), starting time constant of water (T_W) of a hydro generating plant, and governor time constant (T_{GT}) of thermal generating plant. These parameters are varied by ±50% from their nominal values. In this scenario, any one of the considered parameters is changed by ±50% at a time, while maintaining others to their nominal values. Therefore, Figures 12.13-12.16 displayed the profiles of Δf_1, Δf_2, and ΔP_{tie} with respect to deviations in load and chosen system parameters. The performance of the proposed SSA tuned CTI-TID controller in terms of settling time and fitness functions are noted in Table 12.6. According to the results shown in Table 12.6 and Figures 12.13-12.16, the impact of loading and parametric changes on system performance are minor, indicating the robustness and stability of the proposed SSA tuned CTI-TID controller.

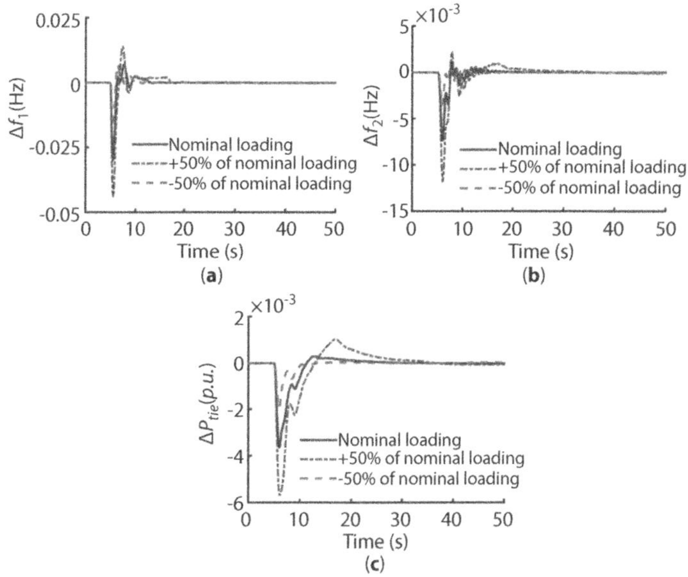

Figure 12.13 Transient response profiles to the ±50% variation in nominal loading condition of Scenario 3 (a) Δf_1 (b) Δf_2 (c) ΔP_{tie}.

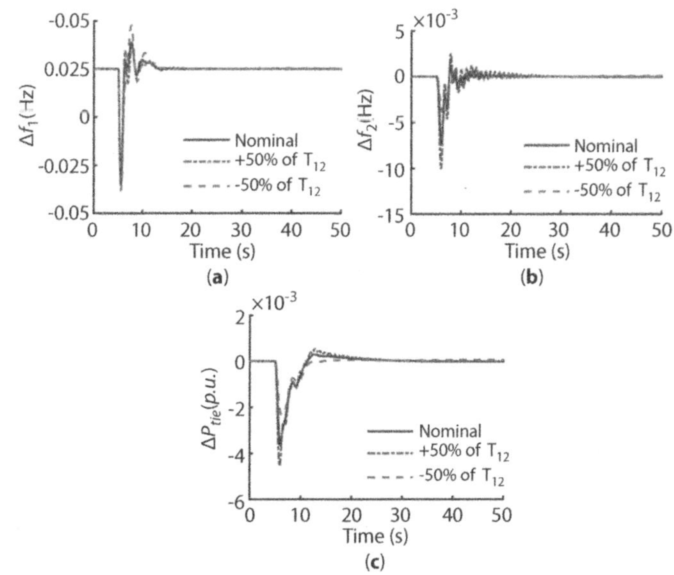

Figure 12.14 Transient response profiles to the ±50% variation in T_{12} of Scenario 3 (a) Δf_1 (b) Δf_2 (c) ΔP_{tie}.

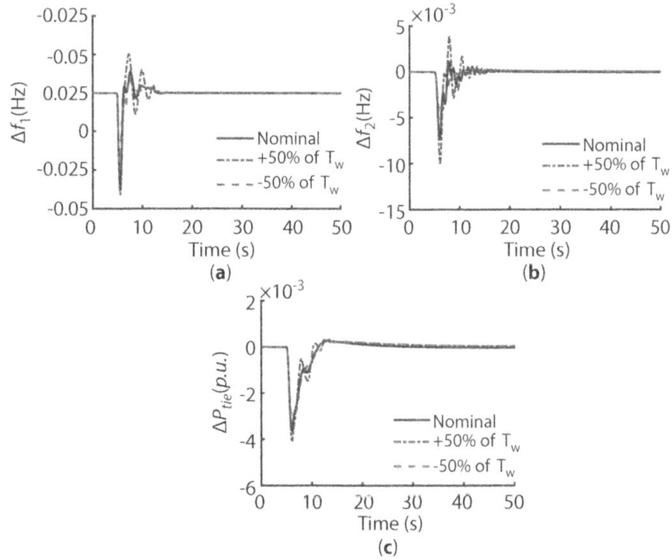

Figure 12.15 Transient response profiles to the ±50% variation in T_W of Scenario 3 (a) Δf_1 (b) Δf_2 (c) ΔP_{tie}.

Table 12.6 Sensitivity analysis of concerned two-areas MSIPS with EVs of Scenario 3.

Parameter variations	% Change	Performance indices				T_s (in s)		
		ITSE ($\times 10^{-2}$)	ISE ($\times 10^{-4}$)	ITAE	IAE ($\times 10^{-2}$)	Δf_1	Δf_2	ΔP_{tie}
Nominal	0	0.24	4.03	0.43	5.18	12.54	17.45	24.52
Loading	+50	0.65	11.0	1.11	10.7	17.04	26.57	34.45
Condition	-50	0.07	1.22	0.22	2.63	10.27	16.48	13.11
T_{12}	+50	0.24	3.95	0.54	5.87	12.80	24.27	25.72
	-50	0.29	4.83	0.51	5.56	12.08	18.54	14.59
T_W	+50	0.40	6.32	0.75	7.72	12.65	16.32	26.09
	-50	0.17	2.79	0.54	5.15	12.64	17.87	28.22
T_{GT}	+50	0.24	4.05	0.57	5.71	12.48	34.89	23.55
	-50	0.24	4.00	0.45	5.26	12.65	13.86	26.07

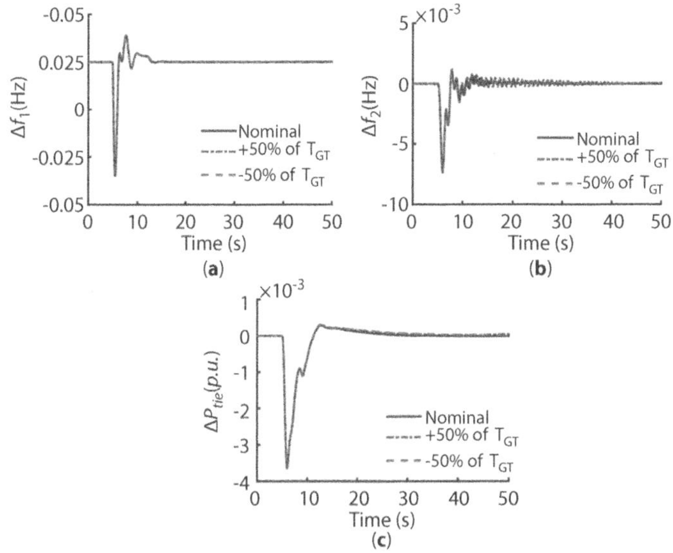

Figure 12.16 Transient response profiles to the ±50% variation in T_{GT} of Scenario 3 (a) Δf_1 (b) Δf_2 (c) ΔP_{tie}.

12.7 Conclusion

In this work, a two-area MSIPS with nonlinearities is attempted to examine the effectiveness of the proposed SSA tuned CTI-TID controller in the presence of EVs. Initially, the tuning efficiency of SSA is tested over TLBO and WCA. Thereafter, the supremacy of an SSA tuned CTI-TID controller is investigated over the other adopted controllers, i.e., SSA tuned CTI-TID controller-based results are compared to the response yielded by the studied SSA tuned CPI-TD controller, SSA tuned PID controller, and other reported IPSO tuned I and PID controllers. Further, the effect of EVs on the studied two-area MSIPS using the proposed SSA tuned CTI-TID is analysed. Comparative analysis of an SSA tuned CTI-TID controller with and without EVs reveals that system dynamics are enhanced with EVs. Finally, a sensitivity analysis is done to demonstrate the designed SSA tuned CTI-TID controller's robustness.

References

1. Kundur P. (1994) *Power system stability and control.* New York, McGraw Hill.
2. Wood AJ, Woolenberg BF. (1998). *Power generation operation and control.* John Wiley & Sons.
3. Jaleeli, N., VanSlyck, L. S., Ewart, D. N., Fink, L. H., & Hoffmann, A. G. (1992). Understanding automatic generation control. *IEEE transactions on power systems, 7*(3), 1106-1122.
4. Shayeghi, H. A. S. H., Shayanfar, H. A., & Jalili, A. (2009). Load frequency control strategies: A state-of-the-art survey for the researcher. *Energy Conversion and management, 50*(2), 344-353.
5. Nanda, J., Mishra, S. and Saikia, L.C. (2009). Maiden application of bacterial foraging-based optimization technique in multi-area automatic generation control. *IEEE Transactions on power systems, 24*(2), pp.602-609.
6. Ray, P. K., Mohanty, S. R., & Kishor, N. (2011). Proportional–integral controller based small-signal analysis of hybrid distributed generation systems. *Energy Conversion and Management, 52*(4), 1943-1954.
7. Chatterjee, K. (2010). Design of dual mode PI controller for load frequency control. *International Journal of Emerging Electric Power Systems, 11*(4).
8. Khodabakhshian, A., & Hooshmand, R. (2010). A new PID controller design for automatic generation control of hydro power systems. *International Journal of Electrical Power & Energy Systems, 32*(5), 375-382.
9. Shabani, H., Vahidi, B., & Ebrahimpour, M. (2013). A robust PID controller based on imperialist competitive algorithm for load-frequency control of power systems. *ISA transactions, 52*(1), 88-95.
10. Saikia, L. C., Nanda, J., & Mishra, S. (2011). Performance comparison of several classical controllers in AGC for multi-area interconnected thermal system. *International Journal of Electrical Power & Energy Systems, 33*(3), 394-401.
11. Franks, R. G., & Worley, C. W. (1956). Quantitative analysis of cascade control. *Industrial & Engineering Chemistry, 48*(6), 1074-1079.
12. Jeng, J. C., & Liao, S. J. (2013). A simultaneous tuning method for cascade control systems based on direct use of plant data. *Industrial & Engineering Chemistry Research, 52*(47), 16820-16831.
13. Dash, P., Saikia, L. C., & Sinha, N. (2016). Flower pollination algorithm optimized PI-PD cascade controller in automatic generation control of a multi-area power system. *International Journal of Electrical Power & Energy Systems, 82*, 19-28.
14. Padhy, S., Panda, S., & Mahapatra, S. (2017). A modified GWO technique based cascade PI-PD controller for AGC of power systems in presence of plug in electric vehicles. *Engineering Science and Technology, an International Journal, 20*(2), 427-442.
15. Tripathy, D., Sahu, B. K., Dev Choudhury, N. B., & Dawn, S. (2018). Spider Monkey optimization based cascade controller for LFC of a hybrid power system. *International Journal of Computational Intelligence & IoT, 2*(4).

16. Dash, P., Saikia, L. C., & Sinha, N. (2015). Automatic generation control of multi area thermal system using Bat algorithm optimized PD–PID cascade controller. *International Journal of Electrical Power & Energy Systems, 68*, 364-372.
17. Jena, T., Debnath, M. K., Padhi, J. R., & Sanyal, S. K. (2018). Load Frequency Control Investigation with Optimum PD-PID Controller with Derivative Filter. *International Journal of Engineering & Technology, 7*(4.39), 117-122.
18. Topno, P. N., & Chanana, S. (2018). Differential evolution algorithm based tilt integral derivative control for LFC problem of an interconnected hydro-thermal power system. *Journal of Vibration and Control, 24*(17), 3952-3973.
19. Debbarma, S., Saikia, L. C., & Sinha, N. (2014). Solution to automatic generation control problem using firefly algorithm optimized IλDμ controller. *ISA transactions, 53*(2), 358-366.
20. Morsali, J., Zare, K., & Hagh, M. T. (2018). Comparative performance evaluation of fractional order controllers in LFC of two-area diverse-unit power system with considering GDB and GRC effects. *Journal of electrical systems and Information Technology, 5*(3), 708-722.
21. Zamani, A., Barakati, S. M., & Yousofi-Darmian, S. (2016). Design of a fractional order PID controller using GBMO algorithm for load–frequency control with governor saturation consideration. *ISA transactions, 64*, 56-66.
22. Latif, A., Das, D. C., Ranjan, S., & Barik, A. K. (2019). Comparative performance evaluation of WCA-optimized non-integer controller employed with WPG–DSPG–PHEV based isolated two-area interconnected microgrid system. *IET Renewable Power Generation, 13*(5), 725-736.
23. Latif, A., Das, D. C., Barik, A. K., & Ranjan, S. (2019). Maiden coordinated load frequency control strategy for ST-AWEC-GEC-BDDG-based independent three-area interconnected microgrid system with the combined effect of diverse energy storage and DC link using BOA-optimized PFOID controller. *IET Renewable Power Generation, 13*(14), 2634-2646.
24. Tasnin, W., & Saikia, L. C. (2018). Maiden application of an sine–cosine algorithm optimized FO cascade controller in automatic generation control of multi-area thermal system incorporating dish-Stirling solar and geothermal power plants. *IET Renewable Power Generation, 12*(5), 585-597.
25. Nayak, J. R., & Shaw, B. (2018). Application of group hunting search optimized cascade PD-fractional order PID controller in interconnected thermal power system. *Trends in Renewable Energy, 4*(3), 22-33.
26. Guha, D., Roy, P. K., & Banerjee, S. (2019). Maiden application of SSA-optimized CC-TID controller for load frequency control of power systems. *IET Generation, Transmission & Distribution, 13*(7), 1110-1120.
27. Ram Babu, N., & Saikia, L. C. (2020). AGC of a multiarea system incorporating accurate HVDC and precise wind turbine systems. *International Transactions on Electrical Energy Systems, 30*(4), e12277.
28. Saha, A., & Saikia, L. C. (2019). Renewable energy source-based multiarea AGC system with integration of EV utilizing cascade controller considering

time delay. *International Transactions on Electrical Energy Systems, 29*(1), e2646.

29. Kumari, S., & Shankar, G. (2018). Novel application of integral-tilt-derivative controller for performance evaluation of load frequency control of interconnected power system. *IET Generation, Transmission & Distribution, 12*(14), 3550-3560.
30. Kumari, S., & Shankar, G. (2019). Maiden application of cascade tilt-integral–tilt-derivative controller for performance analysis of load frequency control of interconnected multi-source power system. *IET Generation, Transmission & Distribution, 13*(23), 5326-5338.
31. Saha, A., & Saikia, L. C. (2018). Performance analysis of combination of ultra-capacitor and superconducting magnetic energy storage in a thermal-gas AGC system with utilization of whale optimization algorithm optimized cascade controller. *Journal of Renewable and Sustainable Energy, 10*(1), 014103.
32. Fan, H., Jiang, L., Zhang, C. K., & Mao, C. (2016). Frequency regulation of multi-area power systems with plug-in electric vehicles considering communication delays. *IET Generation, Transmission & Distribution, 10*(14), 3481-3491.
33. Shimizu, K., Masuta, T., Ota, Y., & Yokoyama, A. (2011, August). A new load frequency control method in power system using vehicle-to-grid system considering users' convenience. In *Proceedings of the 17th Power System Computation Conference, Stockholm, Sweden* (pp. 22-26).
34. Pham, T. N., & Trinh, H. (2015). Load frequency control of power systems with electric vehicles and diverse transmission links using distributed functional observers. *IEEE Transactions on Smart Grid, 7*(1), 238-252.
35. Mirjalili, S., Gandomi, A. H., Mirjalili, S. Z., Saremi, S., Faris, H., & Mirjalili, S. M. (2017). Salp Swarm Algorithm: A bio-inspired optimizer for engineering design problems. *Advances in Engineering Software, 114*, 163-191.
36. Kumari, S., & Shankar, G. (2018, December). A novel application of salp swarm algorithm in load frequency control of multi-area power system. In *2018 IEEE International Conference on Power Electronics, Drives and Energy Systems (PEDES)* (pp. 1-5). IEEE.
37. Barisal, A. K. (2015). Comparative performance analysis of teaching learning based optimization for automatic load frequency control of multi-source power systems. *International Journal of Electrical Power & Energy Systems, 66*, 67-77.
38. Morsali, J., Zare, K., & Hagh, M. T. (2016). Performance comparison of TCSC with TCPS and SSSC controllers in AGC of realistic interconnected multi-source power system. *Ain shams engineering journal, 7*(1), 143-158.
39. Valerio, D., & Da Costa, J. S. (2004). Ninteger: a non-integer control toolbox for MatLab. *Proceedings of fractional differentiation and its applications, Bordeaux.*

Appendix

The parameters of two-area MSIPS model are taken from work of [20] and EVs [30, 33] are as follows:

$P_{r1} = P_{r2} = 2000\text{MW}; P_L = 1840\text{MW} : K_{ps_1} = K_{ps_2} = 68.9566\text{Hz/p.u.MW}; T_{ps_1} = T_{ps_2} = 11.49\text{s};$
$R_T = R_H = R_G = 2.4\text{Hz/p.u.MW}; f = 60Hz; T_{12} = 0.0433; a_{12} = -1; B_1 = B_2 = 0.4312\text{p.u.MW/Hz};$
$T_t = 0.3\text{s};\ T_{GT} = 0.08\text{s};\ T_r = 10\text{s};\ K_r = 0.3;\ T_{RH} = 28.75\text{s};\ T_W = 1\text{s};\ T_{GH} = 0.2\text{s};\ T_{RS} = 5\text{s};\ X_c = 0.6\text{s};\ Y_c = 1\text{s};$
$C_g = 1;\ T_F = 0.23\text{s};\ B_g = 0.05\text{s};\ T_{CD} = 0.2\text{s};\ T_{CR} = 0.01\text{s};\ K_T = 0.543478;\ K_H = 0.326084;\ K_G = 0.130438;$
$N_{control-in}(t) = 12; N_{initial}(t) = 90; N_{control}(t) = 82; N_{plug-out}(t) = 20$ and Number of LC $= 600$.

13

Cyber Security Control Systems for Operational Technology

S. Sriram

Department of Computer Science and Engineering, Bannari Amman Institute of Technology, Sathy - Bhavani State Highway, Alathukombai - Post, Sathyamangalam, Erode District, Tamil Nadu, India

Abstract

Digital related occurrence reaction plans stretch out their network safety projects to address OT [Operational Technology] frameworks. Numerous organizations have insufficient OT digital protection and do not have the controls that are normal spots in IT [Information Technology] organizations, like designing the board, emergency courses of action, access control strategy, and validation strategy. OT framework administrators are essentially worried about the accessibility and dependability of the framework and less so with the security in light of the fact that, verifiably, OT frameworks have been confined from IT organizations. The ascent of computerized advancements brings another degree of digital intricacy to manufacturing plants. The Fourth Industrial Revolution proclaims a period of enormous potential for advancement and development. It additionally brings new dangers and difficulties. Cycle should cover the danger evaluation, security design, and consistence reviews. Innovation should cover control and noteworthy knowledge. Data Technology (IT) and Operational Technology (OT) networks that control tasks progressively to control framework items are alluded to as framework Control Systems, Energy Management Systems, and Manufacturing Execution Systems depending upon the business. There has been a lot of conversation and discussion around the assembly mediating Information Technology (IT) and Operational Technology (OT). With an end goal to give better performance and data stream between income creating OT resources and endeavor applications, these frameworks have frequently been interconnected as a rule without appropriately protecting the control frameworks from digital assault first. On the off chance that the IT and OT networks are interconnected,

Email: sriram@srptc.ac.in; srsriramengg1@gmail.com

Vipin Chandra Pal, Suman Lata Tripathi, and Souvik Ganguli (eds.) Industrial Control Systems, (285–302) © 2024 Scrivener Publishing LLC

yet not appropriately protected, a break to one organization can undoubtedly cross over to the next, leaving the whole processing foundation in danger. IT and OT networks seem to have comparative advances and the digital protection approach is carried out for security reasons. Numerous significant contrasts in IT and OT organizations will be uncovered. The novel qualities of OT frameworks and organizations block numerous conventional IT endeavor security items from working securely without debilitating activities.

***Keywords*:** Operational technology [OT] security, style of cyber security, operational technology security risk, IT & OT networks

13.1 Introduction

Running and working a business implies you have a ton of resources to ensure representatives, cash, and merchandise and the prosperity of your business rides on giving the best insight to your shoppers. In any case, the danger of criminals, hoodlums, and lying clients might put your business in danger [1–5]. At the point when you need to ensure your business and its resources, using incorporated security framework will assist with considering every contingency.

Coordinated security frameworks are the most ideal alternatives for any measured business. In addition to the fact that they offer the best in improvement, they will guarantee every one of your bases are covered. Incorporated security frameworks are a sort of complex framework that utilizes various parts and strings them into one. For a long time, modern frameworks depended upon restrictive conventions and programming that were physically overseen and observed by people and had no association with the rest of the world. They were a genuinely unimportant objective for programmers as there was no organized interface to assault and nothing to acquire. The most ideal way of infiltrating these structures was to get real induction to a terminal and this was no basic task. IT and OT incorporated pretty much nothing and did not manage similar sorts of weaknesses [2, 3].

Modern frameworks were brought online to convey huge information and shrewd investigation, as well as receive new abilities through innovative incorporations. IT and OT union gives associations solitary perspective on modern frameworks along with measurement of the board arrangements that guarantee precise data is conveyed to individual network devices, sensors, and gadgets in the best configuration [8, 14].

When IT and OT frameworks work together, efficiencies are found, frameworks can be distantly observed, and associations can understand security benefits that are utilized on managerial IT frameworks. This progress from shut to open frameworks has produced a huge number of security hazards that should be tended to [17].

13.2 Operational Technology Security Risk

As modern frameworks become more associated, they additionally become more susceptible to weaknesses. The significant expense of modern gear and the destruction to networks and economies that an attack could make are key factors for affiliations wanting to guarantee their modern organizations. Adding legacy equipment and security guidelines that might forbid any alterations made to hardware and consistence guidelines that require touchy information to be made accessible to outsiders. Fortunately, it is feasible to protect modern organizations without disturbing activities or gambling rebelliousness [10]. By utilizing arrangements that permit total performance of an organization and setting up the right security approaches, you can set up a compelling OT system that will ensure your cycles and individuals and benefit and altogether diminish security weaknesses and occurrences [5].

13.2.1 Today's Security of Industrial Networks

Modern conditions generally have lower signal traffic than IT conditions. A significant part of the traffic signal is going between resolved endpoints and subsequently can be base lined and stacked high more effectively than traffic that is produced on an IT organization. Using checking and examination devices can help with recognizing and guarantee against unapproved changes and various irregularities that could signal an attack in full headway or in its fundamental stages. Security is centered around dealing with the insider hazard inside an association and does this through understanding the practices of representatives. With more representatives moving to far off workplaces, the danger from insiders to an association's touchy information has never been more prominent. These vulnerable sides exhibit the requirement for a human-driven and robotized way to deal with network safety to assist with ensuring the two individuals and delicate organization information. These dangers have customarily been overseen through a system of setting up fundamental shields to keep dangers out, as opposed to searching for hazards inside an association [3, 4].

13.2.2 User Activity Monitoring

The greatest danger to your touchy information comes from inside User Activity Monitoring (UAM), which stops dangers the second they are distinguished. Hazardous insiders, regardless of whether malignant or careless, are liable for almost 60% of all information penetrates. With the mass movement to unaided far off workplaces, the danger to an association's delicate information has never been more prominent. Any association with restrictive information or protected innovation is defenseless against insider dangers at the human point, where information is generally important and helpless. In any case, there are approaches to guard your basic information and customer IP while staying away from substantial fines and loss of income and notoriety. Representatives inside the association can evade security arrangements to cause damage or, all the more regularly, accidentally make moves that compromise even the most noteworthy outer danger protections. There are three threat actor types:

- Compromised: It can allude to innovation (e.g., a particular gadget has been compromised frequently without the client's full information) or to a person.
- Malicious: Includes both planned demonstrations of treachery for reasons for monetary benefit or reprisal and purposeful conduct that prompts a break.
- Accidental: Characterized as accidental representative remissness that can uncover your network from messaging client information to some unacceptable beneficiary, to being hoodwinked by the most recent phishing trick.

13.2.3 Hazard in Reputed Industries

The harms from information penetration can be huge for any association and overseeing insider hazard is particularly basic in profoundly managed businesses like healthcare and financial services administrations. Meeting the necessities of administrative norms, including GDPR, HIPAA, ISO 27001, NIST, Fed Ramp, FISMA, and others, can be a mind boggling task in the present developing scene. The expanding speed of information availability and an increment in distant labor forces is leaving associations with vulnerable sides and expected issues with consistence. An insider hazard program ought to keep up with the respectability of information security and protection abilities, while likewise sticking to provincial and industry consistence rules. The results of not overseeing insider dangers can be expensive, with occurrences including careless representatives or project workers costing a normal of $306,110.1 For insider hazards, the board implements consistence, yet in addition mitigates the danger of fines, suit expenses, and harm to an association's standing [5].

13.2.4 Dynamic Security Battle Space

Quite recently, security was a lot of something very similar. If your organization was wise to normal security, you were possibly safe since another person would get hit first. Prevention techniques are designed to block an attacker from getting in, and can be critical to your security strategy is shown in Figure 13.1. Programmers were searching for the "easy breaches", the simple penetrates. Online lawbreakers, for the most part, utilized a wide "shower and implore" way to deal with sharply discovered targets. Back then, "signature-based" arrangements which lead to a distinguished known, pernicious program and squared them, appeared well and good. In the event that one organization saw another danger, a mark could be composed for it and dispersed to others to shield them from contamination. Quick forward to now: as associations have supported their security, programmers likewise advanced. Assaults today are more refined and focused than any other time in recent memory. Rather than sending nonexclusive malware, programmers

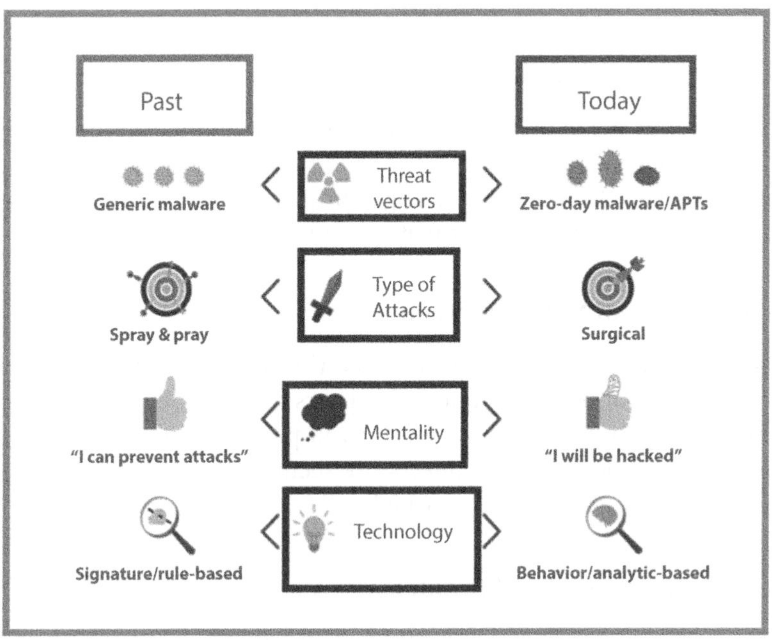

Figure 13.1 Prevention is not enough based on cyber security.

today cautiously plot every single assault, utilizing special "zero-day" misuses that render signature-based assurances almost futile.

13.3 Taxonomy of Security Vulnerabilities

13.3.1 Buffer Overflow

This weakness happens when information is composed past the constraints of a cradle. Cushions are memory regions assigned to a function. By altering the information past the limits of a cushion, the function gets memory assigned for different cycles. It can prompt a framework crash, data compromise, or give heightening advantages.

13.3.2 Non-Substantial Input

Projects frequently function with information. The data program might be a malevolent substance intended to constrain a program to act in an accidental manner. The program then gets a picture for preparing. A noxious client can make a picture document with invalid picture measurements. The noxiously created measurements could constrain the program to allot supports of wrong and startling sizes.

13.3.3 Race Conditions

This weakness is the point at which the yield of an occasion relies upon requested or planned yields. A race condition turns into a wellspring of weakness when the necessary arranged or planned occasions do not happen in the right request or legitimate planning.

13.3.4 Lack of Security Practices

Frameworks information can be ensured through procedures like confirmation, approval, and encryption. Designers ought not endeavor to make their own security calculations since it will probably present weaknesses. It is firmly prompted that engineers use security libraries that have been effectively made, tried, and confirmed.

13.3.5 Access Control Problems

It is the way to control actual admittance to gears, directing the approaches of an asset, like a document, and how they can manage it, for example, read the record. Numerous weaknesses are made by ill-advised utilization of controls. Practical entrance security can be survived if the aggressor has actual admittance to target. For instance, regardless of what you set a record's authorizations to, the working framework cannot keep somebody from bypassing the working framework and perusing the information circle to get data protection.

13.3.6 Malicious Software

This is program code that can be used to take data processes that impact harm to the system or design of a structure.

13.3.7 Spyware

Spyware is designed to track and watch out for customers. It oftentimes joins activity trackers, keystroke grouping, and data mining. While attempting to overcome wellbeing endeavors, spyware oftentimes changes security protection.

13.3.8 Program in Adware

Promoting upheld programming may be intended to naturally convey notices. It is regularly introduced for certain forms of programming. Some adware is intended to just convey notices yet it is likewise entirely expected for adware to accompany spyware.

13.3.9 Bot

A bot is malware intended to consequently perform an activity, normally on the web. A few PCs are contaminated with bots modified to discreetly sit tight for orders given by the attacker.

13.3.10 Ransomware

This malware in a PC framework takes information hostage until an installment is done. Ransomware, for the most part, works by scrambling information in the PC with a critical obscurity to the client. Some different adaptations of ransomware can exploit explicit framework weaknesses to secure the framework. Ransomware is covered by a downloaded record or some product weakness.

13.3.11 Scareware

This malware is intended to convince a system to make a particular move dependent on dread. It manufactures spring up windows that look like working framework exchange windows. These windows pass on manufactured messages expressing the framework is in danger or necessitates a process of program execution to get back to typical activity.

13.3.12 Rootkit

It is a kind of malware intended to adjust the working framework to make a secondary passage. Assailants then utilize the secondary passage to get to the PC distantly. Most rootkits exploit programming weaknesses to perform advantage heightening and adjust framework documents. It is likewise entirely expected for rootkits to alter framework criminology and observing apparatuses, making them extremely difficult to identify.

13.3.13 Virus

An infection of pernicious code connected to documents and frequently, authentic projects. Most infections require end-client enactment and can actuate at a particular time or date. Infections can be innocuous and basically show an image or they can be ruinous, for example, those that alter or erase information. Infections can likewise be customized to transform to keep away from discovery. Most infections are presently spread by peripheral devices or email.

13.3.14 Trojan Horse

It is malware whose presence is under the appearance of an ideal activity. This noxious code misuses the advantages of the client that runs it. Regularly, Trojans are found in picture documents, sound records, or games. It contrasts from an infection since it ties itself to non-executable records.

13.3.15 Worms

Worms are malicious codes that repeat themselves by autonomously abusing weaknesses in networks. Worms typically hinder networks. Though an infection requires a host program to run, worms can run without help from anyone else. Other than the underlying contamination, they presently do not need client interest. The worm can spread rapidly over the organization. It shares comparative examples. Worms have an empowering weakness, an approach to engender themselves, and they all contain a payload.

13.3.16 Man-In-The-Middle [MitM]

It permits the assailant to assume responsibility for gadgets, except the client's information. With that degree of access, the assailant can catch client data prior to handing-off it to its expected objective. MitM assaults are generally used to take monetary data. Numerous malware and procedures exist to give assailants MitM capacities. Despite the kind of malware, the framework is contaminated with, malware indications include:

- There is an increase in CPU usage and decrease in computer speed.
- The computer freezes or crashes often.
- The content files are modified and deleted.

13.3.17 Blended Attacks

Mixed assaults are assaults that utilize numerous methods to think twice about a target by utilizing a few diverse assault methods immediately, malware that are a mixture of worms, Trojan horses, spyware, keyloggers, spam, and phishing plans. This pattern of mixed assaults is uncovering more intricate malware and putting client information at incredible danger.

13.4 Methodology

13.4.1 Stronger Operational Technology [OT] Security

OT security is the act of utilizing equipment and programming advances to screen, distinguish, and control changes to cycles, occasions, and gadgets. The reason behind utilizing OT security is to ensure mechanical frameworks and organizations, for example, savvy city machines and transportation organizations.

13.4.2 Creating Inventory and Identifying OT Vulnerabilities

OT specialists are needed to build up a precise OT resource stock with baselines for each. A total organization map is additionally needed to plan all inbound and outbound communication. A total appraisal ought to be made to distinguish weakness to OT resources and security controls needed to moderate those dangers.

13.4.3 Acquiring Automated Threat Intelligence Feeds

Danger insight information gives significant data with respect to dangers. Danger insight takes care of accessibility, including industry, administrative, and business tasks [6]. Be that as it may, one of the significant concerns is changing over such insight, which takes care of noteworthy knowledge. To this end, OT security experts ought to send mechanized danger ingestion abilities in network observing arrangements.

13.4.4 Back/Restore

Creating an optimal backup of OT data is a vital approach that can ensure data availability even after a data breach. To this end, first, create a back copy of the OT data and then perform a test restore to ensure that the entire backup system and restore system is working properly and accurately [7].

13.5 Style of Cyber Security

13.5.1 Security Automation

The capacity to consequently coordinate Cyber Threat Intelligence into your security framework for proactive insurance is called "Robotized DNS Sinkhole Breach Detection" and security mechanization abilities by locally incorporating Cyber Threat Intelligence into "Computerized Penetration Testing" and "Robotized Incident Response" answers for improved break identification and examination. Sharing Cyber Threat Intelligence, both secretly and openly, would now be able to be performed inside a couple of snaps. Your business and your industry accomplices can now naturally acquire the advantages of industry-explicit Cyber Threat Intelligence through robotized sharing and reconciliation with your security answers for proactive relevant break locations and anticipation [9]. It naturally expands the insight of your association to assist with guaranteeing that you keep steady over the most recent dangers, assaults, and security breaks to protect your business.

13.5.2 Breach Detection System (BDS)

Breach detection tools can help to identify threats inside your network. They are either software or hardware products capable of recognising active threats and alerting relevant security staff that they need to take action as shown in Figure 13.2. Break location frameworks classify the uses and security gadgets intended to distinguish the movement of present malware in an organization after a penetration has happened. Enterprise IT utilizes BDS to ensure against the assortment of cutting edge dangers, particularly unidentified malware. In contrast to level 1 security, like a firewall or interruption anticipation, that check approaching traffic, BDS centers around malevolent movement inside the organization it ensures. It decides potential penetrates by varying mixes of heuristics, traffic investigation, hazard evaluation, safe stamped traffic, information strategy comprehension, and infringement detailing. Utilizing these techniques, BDS can now and again discover penetrates as they happen and at different occasions distinguish breaks and side-channel assaults that had not recently been found. BDS assists with

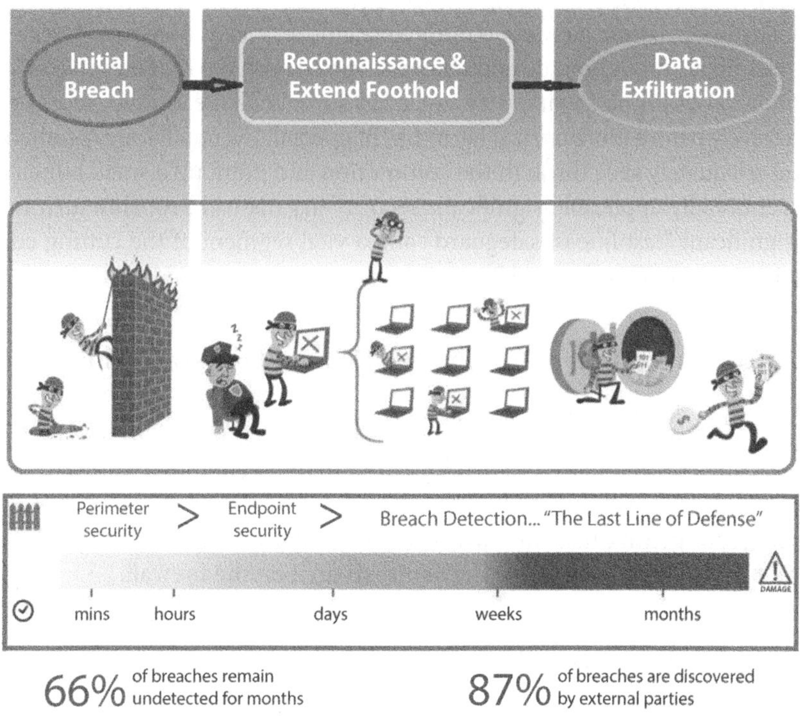

Figure 13.2 Breach detection.

tracking down the obscure progressed and versatile dangers. Indeed, even significant sites have been hacked and moreover, the normal effective break endures 16 months. On the two checks, there is surely space to eliminate harms. The utilization of BDS address a change in the way of thinking from forestalling each interruption to understanding that interruptions will occur and zeroing in on getting those interruptions sooner. BDS should be designed with subtleties like working framework, a rundown of supported applications, and projects permitted to interface with the Internet.

This load of elements highlights the requirement for hearty break identification to give a "last line of safeguard" against these assaults, rather than simply zeroing in on hindering the underlying influx of the interruption. While afterward location is certainly not another idea, generation of (IDS) Intrusion Detection Systems and Security Information and Event Management (SIEM) technologies generally fall short in today's data-center environment. The primary explanation is that they depend on pre-characterized rules/marks to identify penetrates. In a universe of bespoke assaults, such arrangements will regularly miss the most applicable alarms, low sign to-commotion. Envision being overflowed with hundreds or an even larger number of alarms each day with everyone addressing a likely penetrate to your association's most touchy information. New penetration recognition means new businesses can detect these developments and changes. Joined with a cozy comprehension of how programmers work, they can at last sort out all the unique pieces progressively before more critical harm has happened. Eventually, arrangements that adequately slice through the commotion and point up a small bunch of exceptionally applicable significant security alarms will probably turn into a significant "last line of safeguard" and a vital segment of the cutting edge venture security stack [11–13].

13.5.3 Protection of Computing Devices From Intrusion

13.5.3.1 Keep the Firewall on Condition

Regardless of whether it is a product firewall or an equipment firewall on a switch, the firewall ought to be turned on and refreshed to keep programmers from getting to your own or organization information. Windows 7 and 8.1 or Windows 10 have the ability to turn on the firewall.

13.5.3.2 Antivirus and Antispyware

Trojan horses, worms, ransomware, and spyware are introduced on your registered gadgets without your authorization to access your PC and your information. Infections can annihilate your information, hinder your PC, or assume control over your PC. One way infections can assume control over your PC is by permitting spammers to communicate messages utilizing your record. Spyware can screen your online exercises, gather your own data, or produce undesirable spring up advertisements on your internet browser while you are on the web. A decent principle is to just download programming from believed sites to try not to get spyware in any case. Antivirus programming is intended to check your PC and approaching emails for infections and erase them. Here and there antivirus programming additionally incorporates antispyware. Stay up to date with the latest to shield your PC from the most up to date malicious software [13].

13.5.3.3 Manage Your Operating System and Browser

Programmers are continually attempting to exploit weaknesses in your working frameworks and your internet browsers. To ensure your PC and information, set the security settings on your PC and program at medium or higher. Update your PC's working framework including your internet browsers and consistently download and introduce the most recent programming patches and security refreshes from the sellers [15].

13.5.3.4 Protection of Smart Devices

Your registered gadgets, regardless of whether they are PCs, workstations, tablets, or advanced cells, ought to be secret word secured to forestall unapproved access. The put away data ought to be scrambled, particularly for touchy or private information. For cell phones, store essential data in the event that these gadgets are taken or lost when you are away from your home. On the off chance that any of your gadgets is compromised, the lawbreakers might access all of your information through your distributed storage specialist organization, for example, iCloud or Google Drive.

13.5.3.5 Unique Passwords for Each Online Account

Most people have more than one online records and each record ought to have a special secret phrase. That is a great deal of passwords to recollect. The outcome of not utilizing solid and exceptional passwords leaves you and your information powerless against digital lawbreakers. Utilizing similar secret words for all your online records resembles utilizing a similar key for all your locked entryways and if an aggressor was to get your key, he would get to all that you own. On the off chance that lawbreakers get your secret key through phishing for instance, they will attempt to get into your other online records. In the event that you just utilize one secret phrase for all records, they can get into every one of your records, take or delete all of your information, or choose to mimic you. Online records that need passwords turns out to be an excessive amount to recollect. One answer for abstaining from reusing passwords or utilizing powerless passwords is to utilize a secret word supervisor [16]. A secret word chief stores and encodes the entirety of your unique and complex passwords. The administrator would then be able to assist you with signing into your online records consequently. You just need to recollect your lord secret key to get to the secret phrase director and deal with the entirety of your records and passwords.

13.5.3.6 Detecting Attacks in Real Time

Programming is not awesome. At the point when a programmer abuses a blemish in a piece of programming before the maker can fix it, it is known as a zero-day assault. Because of the complexity and immensity of zero-day assaults discovered today, it is turning out to be entirely expected that organization assaults will succeed and that a fruitful safeguard is currently estimated in how rapidly an organization can react to an assault. The capacity to identify assaults as they occur continuously, just as halting the assaults promptly or not long after happening, is the best objective. Lamentably, numerous organizations and associations today cannot identify assaults until days or even a very long time after they have happened. Constant scanning from Edge to Endpoint used for detecting assaults progressively requires effectively examining for assaults utilizing firewall and IDS/IPS network gadgets. Cutting edge customer/worker malware discovery with associations with online worldwide danger communities should likewise be utilized [18–20]. Today, dynamic examining gadgets and programming should distinguish network inconsistencies utilizing settings based investigation and conduct recognition.

13.5.3.7 Cyber Attacks in Operational Technology

Security fixing doesn't work the same way in OT. Since patching OT components requires complete shutdowns that halts production, vendors running OT networks rarely patch their components, if at all. Since OT

Table 13.1 Cyber attacks 2020.

Cyber attacks	Organization
Software AG affected by a ransomware attack in October 2020	**Software AG:** Germany
Sopra Steria French IT service hit by ransomware attack	**Sopra Steria**: French IT service giant
Seyfarth Shaw LLP hit by an "aggressive malware" attack	**Seyfarth Shaw LLP**: Global legal firm
Carnival Corporation reported a data breach hit by ransomware attack	**Carnival Corporation**: Cruise line operator

parts are seldom refreshed, they might have a lot more open weaknesses when contrasted with IT PCs. This implies that the likelihood of a fruitful endeavor on an OT framework is dramatically higher than on an IT framework [6, 7].

A cyber attacks refers to an action designed to target a computer or any element of a computerized information system to change, destroy, or steal data, as well as exploit or harm a network. Cyber attacks have been on the rise, in sync with the digitization of business that has become more and more popular in recent years. Table 13.1 represent some glimpse of attack.

13.6 Avoidance of Threads in Operational Technology

13.6.1 [DDoS] Distributed Denial of Services Attacks and Response

DDoS is one of the greatest assault dangers requiring constant reaction and location as shown in Figure 13.3. DDoS assaults are amazingly hard to shield against on the grounds that the assaults begin from hundreds or thousands of zombies and the assaults show up as real traffic, as displayed in the figure. For some organizations and associations, consistent DDoS assaults cripple Internet workers and organization accessibility. The capacity to recognize and react to DDoS assaults continuously is vital.

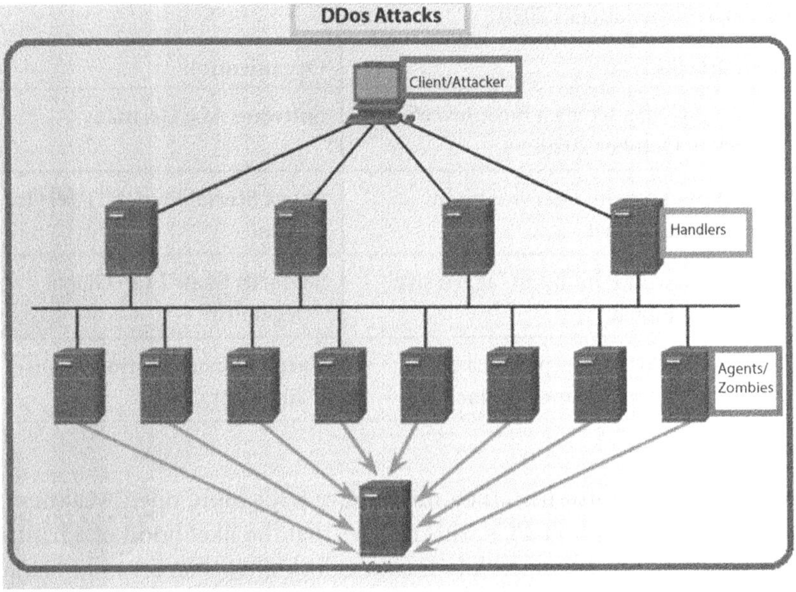

Figure 13.3 Distributed denial of services [DDoS] attacks. DDoS attack aims to make your website and servers unavailable to legitimate users. DDoS can also be used as a smokescreen for other malicious activities and to take down security appliances, breaching the target's security perimeter.

13.6.2 Protecting Against Malware in Operational Technology

The Advanced Malware Protection (AMP) Threat Grid investigates a large number of records and relates them against countless other dissected malware

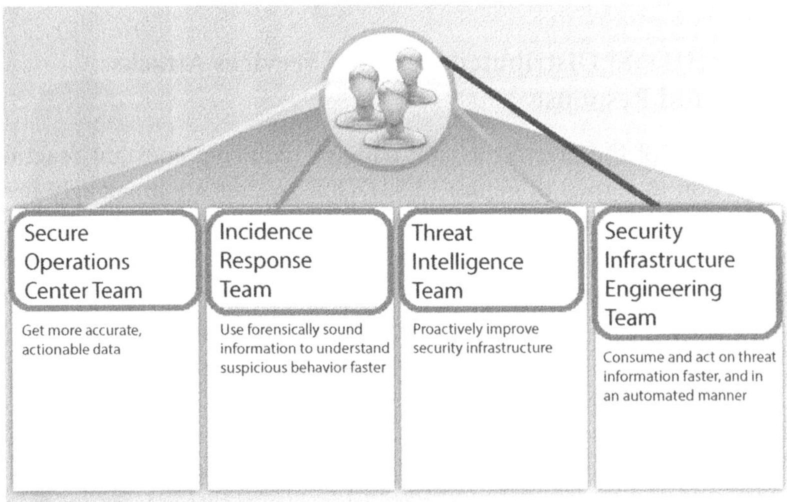

Figure 13.4 Advanced malware protection [AMP] threat grid.

antiquities. AMP Threat Grid Appliances as shown in Figure 13.4 combine advanced malware analysis with comprehensive threat analytics and content in one on-premises appliance. They are designed for organizations that have compliance or policy restrictions on submitting malware samples to the cloud. This gives a worldwide perspective on malware assaults, crusades, and their dispersion. AMP is customer/worker program conveyed on having endpoints, as an independent worker, or on other organization security gadgets.

13.7 Conclusion

Operational Technology [OT] frameworks are generally being utilized in modern conditions and network protection has become their fundamental issue. Nonetheless, you can keep away from and moderate OT dangers by proactively recognizing, characterizing, and checking your OT foundation. To proactively identify, you are required to use a Threat Hunting technique. You can use an automated threat hunting system. The safety efforts ought to be planned and given and initially an organization should know its need of safety on the various levels of the association and afterward it ought to be carried out for various levels. Security approaches ought to be planned first before its execution in such a manner, so future adjustment and selection can be adequate and effectively reasonable. The security framework should be tight, however should be adaptable for the end-client to make him agreeable.

References

1. Von Solms, R., & Van Niekerk, J. (2013). From information security to cyber security. computers & security, 38, 97-102.
2. Murray, G., Johnstone, M. N., & Valli, C. (2017). The convergence of IT and OT in critical infrastructure.
3. Brocklehurst, S., Littlewood, B., Olovsson, T., & Jonsson, E. (1994). On measurement of operational security. IEEE Aerospace and Electronic Systems Magazine, 9(10), 7-16.
4. Dacier, M., Deswarte, Y., & Kaâniche, M. (1996, January). Models and tools for quantitative assessment of operational security. In IFIP International Conference on ICT Systems Security and Privacy Protection (pp. 177-186). Springer, Boston, MA.
5. Van Brabant, K. (2000). Operational security management in violent environments (p. 330). London: Overseas Development Institute.
6. Haddad, S., Dubus, S., Hecker, A., Kanstrén, T., Marquet, B., & Savola, R. (2011, September). Operational security assurance evaluation in open

infrastructures. In 2011 6th International Conference on Risks and Security of Internet and Systems (CRiSIS) (pp. 1-6). IEEE.
7. Cavelty, M. D. (2010). Cyber-security. The routledge handbook of new security studies, 154-162.
8. Sun, C. C., Hahn, A., & Liu, C. C. (2018). Cyber security of a power grid: State-of-the-art. International Journal of Electrical Power & Energy Systems, 99, 45-56.
9. Thakur, K., Qiu, M., Gai, K., & Ali, M. L. (2015, November). An investigation on cyber security threats and security models. In 2015 IEEE 2nd International Conference on Cyber Security and Cloud Computing (pp. 307-311). IEEE.
10. Ullah, F., Naeem, H., Jabbar, S., Khalid, S., Latif, M. A., Al-Turjman, F., & Mostarda, L. (2019). Cyber security threats detection in internet of things using deep learning approach. IEEE Access, 7, 124379-124389.
11. Julisch, K. (2008, September). Security compliance: The next frontier in security research. In Proceedings of the 2008 New Security Paradigms Workshop (pp. 71-74).
12. Goo, J., Yim, M. S., & Kim, D. J. (2014). A path to successful management of employee security compliance: An empirical study of information security climate. IEEE Transactions on Professional Communication, 57(4), 286-308.
13. Safa, N. S., Von Solms, R., & Furnell, S. (2016). Information security policy compliance model in organizations. computers & security, 56, 70-82.
14. Rid, T., & Buchanan, B. (2015). Attributing cyber attacks. Journal of Strategic Studies, 38(1-2), 4-37.
15. Cashell, B., Jackson, W. D., Jickling, M., & Webel, B. (2004). The economic impact of cyber-attacks. Congressional research service documents, CRS RL32331 (Washington DC), 2.
16. Zhu, B., Joseph, A., & Sastry, S. (2011, October). A taxonomy of cyber attacks on SCADA systems. In 2011 International conference on internet of things and 4th international conference on cyber, physical and social computing (pp. 380-388). IEEE.
17. Li, X., Liang, X., Lu, R., Shen, X., Lin, X., & Zhu, H. (2012). Securing smart grid: Cyber attacks, countermeasures, and challenges. IEEE Communications Magazine, 50(8), 38-45.
18. Uma, M., & Padmavathi, G. (2013). A Survey on Various Cyber Attacks and their Classification. IJ Network Security, 15(5), 390-396.
19. Cheung, S., Lindqvist, U., & Fong, M. W. (2003, April). Modeling multistep cyber attacks for scenario recognition. In Proceedings DARPA Information Survivability Conference And Exposition (Vol. 1, pp. 284-292). IEEE.
20. Hideshima, Y., & Koike, H. (2006, January). STARMINE: A visualization system for cyber attacks. In Proceedings of the 2006 Asia-Pacific Symposium on Information Visualisation-Volume 60 (pp. 131-138).

About the Editors

Vipin Chandra Pal, PhD has more than 6 years of experience in teaching and is currently associated with the National Institute of Technology Silchar as an assistant professor. He has acted as a student chair in organizing the conferences at MNNIT Allahabad, as well as several faculty development programs, workshops, and expert lectures for the students. He has published eight Science Citation Index and Scopus research papers and presented seven research papers for national and international conferences. Recently, he has filed five Australian Innovation patents in the area of control systems.

Suman Lata Tripathi, PhD is associated with Lovely Professional University as a professor with more than seventeen years of experience in academics. She has published more than 55 research papers in revered journals and conferences. She has also edited more than 12 books and a book series in different areas of electronics and electrical engineering.

Souvik Ganguli, PhD is associated with the Thapar Institute of Engineering and Technology, Patiala as an assistant professor since June 2009 with fourteen years of experience in academics. Before joining academics he served the industry for more than two years. He has published eight Science Citation Index journal papers and nearly 50 Scopus indexed papers, book chapters, and conferences. Recently, he has been granted an Australian Innovation patent for his contribution to the industrial cyber-physical system and eight of his patents are already published and awaiting grants.

Index

Also of Interest

By the same editors

NANODEVICES FOR INTEGRATED CIRCUIT DESIGN, Edited by Suman Lata Tripathi, Abhishek Kumar, K. Srinivasa Rao, and Prasantha R. Mudimela, ISBN: 9781394185788. Written and edited by a team of experts in the field, this important new volume broadly covers the design of nano-devices and their integrated applications in digital and analog integrated circuits (IC) design.

EXPLAINABLE MACHINE LEARNING MODELS AND ARCHITECTURES: Real-Time System Implementation, Edited by Suman Lata Tripathi and Mufti Mahmud, ISBN: 9781394185849.This cutting-edge new volume covers the hardware architecture implementation, the software implementation approach, and the efficient hardware of machine learning applications.

MACHINE LEARNING TECHNIQUES FOR VLSI CHIP DESIGN, Edited by Abhishek Kumar, Suman Lata Tripathi, and K. Srinivasa Rao, ISBN: 9781119910398. This cutting-edge new volume covers the hardware architecture implementation, the software implementation approach, and the efficient hardware of machine learning applications with FPGA or CMOS circuits, and many other aspects and applications of machine learning techniques for VLSI chip design.

INTELLIGENT GREEN TECHNOLOGIES FOR SMART CITIES, Edited by Suman Lata Tripathi, Souvik Ganguli, Abhishek Kumar, and Tengiz Magradze, ISBN: 9781119816065. Presenting the concepts and fundamentals of smart cities and developing "green" technologies, this volume, written and edited by a global team of experts, also goes into the practical applications that can be utilized across multiple disciplines and industries, for both the engineer and the student.

DESIGN AND DEVELOPMENT OF EFFICIENT ENERGY SYSTEMS, edited by Suman Lata Tripathi, Dushyant Kumar Singh, Sanjeevikumar Padmanaban, and P. Raja, ISBN 9781119761631. Covering the concepts and fundamentals of efficient energy systems, this volume, written and edited by a global team of experts, also goes into the practical applications that can be utilized across multiple industries, for both the engineer and the student.

Electrical and Electronic Devices, Circuits, and Materials: Technical Challenges and Solutions, edited by Suman Lata Tripathi, Parvej Ahmad Alvi, and Umashankar Subramaniam, ISBN 9781119750369. Covering every aspect of the design and improvement needed for solid-state electronic devices and circuit and their reliability issues, this new volume also includes overall system design for all kinds of analog and digital applications and developments in power systems.

Green Energy: Solar Energy, Photovoltaics, and Smart Cities, edited by Suman Lata Tripathi and Sanjeevikumar Padmanaban, ISBN 9781119760764. Covering the concepts and fundamentals of green energy, this volume, written and edited by a global team of experts, also goes into the practical applications that can be utilized across multiple industries, for both the engineer and the student.

Check out these other related titles from Scrivener Publishing

MODELING AND OPTIMIZATION OF OPTICAL COMMUNICATION NETWORKS, Edited by Chandra Singh, Rathishchandra R Gatti, K.V.S.S.S.S. Sairam, and Ashish Singh, ISBN: 9781119839200. *Modeling and Optimization of Optical Communication Networks* is a comprehensive and authoritative book that delves into the optical networks, principles, technologies, and practical applications of optical networks, equipping readers with the knowledge needed to design, implement, and optimize optical networks for various applications, from telecommunications to scientific research. With its comprehensive coverage and up-to-date insights, this book serves as an essential reference in the field of optical networks.

RF CIRCUITS FOR 5G APPLICATIONS: Designing with mm Wave Circuitry, Edited by Sangeeta Singh, Rajeev Kumar Arya, B.C. Sahana and Ajay Kumar Vyas, ISBN: 9781119791928. This book addresses FinFET-based analog IC designing for fifth generation (5G) communication networks and highlights the latest advances, problems, and challenges while presenting the latest research results in the field of mmwave integrated circuits design.

WIRELESS COMMUNICATION SECURITY: Mobile and Network Security Protocols, Edited by Manju Khari, Manisha Bharti, and M. Niranjanamurthy, ISBN: 9781119777144. Presenting the concepts and advances of wireless communication security, this volume, written and edited by a global team of experts, also goes into the practical applications for the engineer, student, and other industry professionals.

ADVANCES IN DATA SCIENCE AND ANALYTICS, edited by M. Niranjanamurthy, Hemant Kumar Gianey, and Amir H. Gandomi, ISBN: 9781119791881. Presenting the concepts and advances of data science and analytics, this volume, written and edited by a global team of experts, also goes into the practical applications that can be utilized across multiple disciplines and industries, for both the engineer and the student, focusing on machining learning, big data, business intelligence, and analytics.

ARTIFICIAL INTELLIGENCE AND DATA MINING IN SECURITY FRAMEWORKS, Edited by Neeraj Bhargava, Ritu Bhargava, Pramod Singh Rathore, and Rashmi Agrawal, ISBN 9781119760405. Written and edited by a team of experts in the field, this outstanding new volume offers solutions to the problems of security, outlining the concepts behind allowing computers to learn from experience and understand the world in terms of a hierarchy of concepts.

MACHINE LEARNING AND DATA SCIENCE: Fundamentals and Applications, Edited by Prateek Agrawal, Charu Gupta, Anand Sharma, Vishu Madaan, and Nisheeth Joshi, ISBN: 9781119775614. Written and edited by a team of experts in the field, this collection of papers reflects the most up-to-date and comprehensive current state of machine learning and data science for industry, government, and academia.

MEDICAL IMAGING, Edited by H. S. Sanjay, and M. Niranjanamurthy, ISBN: 9781119785392. Written and edited by a team of experts in the field, this is the most comprehensive and up-to-date study of and reference for the practical applications of medical imaging for engineers, scientists, students, and medical professionals.

SECURITY ISSUES AND PRIVACY CONCERNS IN INDUSTRY 4.0 APPLICATIONS, Edited by Shibin David, R. S. Anand, V. Jeyakrishnan, and M. Niranjanamurthy, ISBN: 9781119775621. Written and edited by a team of international experts, this is the most comprehensive and up-to-date coverage of the security and privacy issues surrounding Industry 4.0 applications, a must-have for any library.

CYBER SECURITY AND DIGITAL FORENSICS: Challenges and Future Trends, Edited by Mangesh M. Ghonge, Sabyasachi Pramanik, Ramchandra Mangrulkar, and Dac-Nhuong Le, ISBN: 9781119795636. Written and edited by a team of world renowned experts in the field, this groundbreaking new volume covers key technical topics and gives readers a comprehensive understanding of the latest research findings in cyber security and digital forensics.

DEEP LEARNING APPROACHES TO CLOUD SECURITY, edited by Pramod Singh Rathore, Vishal Dutt, Rashmi Agrawal, Satya Murthy Sasubilli, and Srinivasa Rao Swarna, ISBN 9781119760528. Covering one of the most important subjects to our society today, this editorial team delves into solutions taken from evolving deep learning approaches, solutions allow computers to learn from experience and understand the world in terms of a hierarchy of concepts.

MACHINE LEARNING TECHNIQUES AND ANALYTICS FOR CLOUD SECURITY, Edited by Rajdeep Chakraborty, Anupam Ghosh and Jyotsna Kumar Mandal, ISBN: 9781119762256. This book covers new methods, surveys, case studies, and policy with almost all machine learning techniques and analytics for cloud security solutions.

SECURITY DESIGNS FOR THE CLOUD, IOT AND SOCIAL NETWORKING, Edited by Dac-Nhuong Le, Chintin Bhatt and Mani Madhukar, ISBN: 9781119592266. The book provides cutting-edge research that delivers insights into the tools, opportunities, novel strategies, techniques, and challenges for handling security issues in cloud computing, Internet of Things and social networking.

DESIGN AND ANALYSIS OF SECURITY PROTOCOLS FOR COMMUNICATION, Edited by Dinesh Goyal, S. Balamurugan, Sheng-Lung Peng and O.P. Verma, ISBN: 9781119555643. The book combines analysis and comparison of various security protocols such as HTTP, SMTP, RTP, RTCP, FTP, UDP for mobile or multimedia streaming security protocol.

SMART GRIDS FOR SMART CITIES VOLUME 2: Real-Time Applications in Smart Cities, Edited by O.V. Gnana Swathika, K. Karthikeyan, and Sanjeevikumar Padmanaban, ISBN: 9781394215874. Written and edited by a team of experts in the field, this second volume in a two-volume set focuses on an interdisciplinary perspective on the financial, environmental, and other benefits of smart grid technologies and solutions for smart cities.

SMART GRIDS AND INTERNET OF THINGS, Edited by Sanjeevikumar Padmanaban, Jens Bo Holm-Nielsen, Rajesh Kumar Dhanaraj, Malathy Sathyamoorthy, and Balamurugan Balusamy, ISBN: 9781119812449. Written and edited by a team of international professionals, this ground-breaking new volume covers the latest technologies in automation, tracking, energy distribution and consumption of Internet of Things (IoT) devices with smart grids.

Printed and bound by CPI Group (UK) Ltd, Croydon, CR0 4YY

17/09/2026

14964247-0003